Department of the Environment

Acidic Emissions Abatement Processes

Manual of Acidic Emission Abatement Technologies

VOLUME 4:

London: HMSO

© Crown copyright 1991
First published 1991

ISBN 0 11 752216 3

Preface

The work represented in this Volume was completed during the period 1986 to 1987 to provide data for the Department of the Environment and others in consideration and implementation of the European Community Directives on Air Pollution which were then under consideration.

Since this volume was written there has been considerable advances in European legislation. Sections 0.4, 1.3 and 3.4 take no account of the Consolidated Directive 91/441 for cars or of 88/77/EEC and 91/542 for heavy vehicles. Likewise, Section 7 does not cover the latest modifications to the UK in-service test procedure.

The Consolidated Directive 91/441 limits for 1993 onwards will make it more difficult for the true lean-burn engine and the small DI diesel to be introduced into cars and both will need further development, justified by their potential for significantly better fuel economy. Details are as follows.

1. CARS :

1.1 Changes in legislation :

1.1.1 All previous Legislation 70/220/EEC, 74/290/EEC, 77/102/EEC, 78/665/EEC, and 83/361/EEC, grouped cars according to their weight. This has been superseded in 91/441/EEC by a maximum weight of 2840 kg, below which all cars have to meet the new emission limits.

1.1.2 Introduction dates for 91/441/EEC :

1st Jan 1992   No car should be denied EEC Type approval or be prohibited from entry into service, on grounds relating to air pollution by its emissions, if it meets 91/441/EEC Type Approval limits.

1st July 1992   No car should be granted Type Approval unless it meets 91/441/EEC.

31st Dec 1992   Initial entry into service of any vehicle not meeting 91/441/EEC shall be prohibited.

1.1.3 Limits Set (91/441/EEC) :

	CO (g/km)	THC + NOx (g/km)	Pm (g/km)
Type Approval :	2.72	0.97	0.14
Conformity of Production :	3.16	1.13	0.18

1.1.4 Changes in the Test procedure :

- Units for the limits have been changed from (g/test) to (g/km).

- Emission limits now include, for the first time, the mass in g/km of particulates.

- Whereas the previous test comprised four Urban Driving Cycles, the new test has a high speed Extra Urban Driving Cycle (EUDC) added. The overall test procedure is shown in Fig 00.1. It is therefore not possible to compare results from tests performed under the previous legislation 83/361/EEC (or earlier), with results from 91/441/EEC.

1.1.5 Further reduction in limit values :

A proposal to reduce limit values further, taking into account technical progress, will be submitted to the European Council of Ministers before 31st December 1992. Although a decision on this proposal will be made before 31st December 1993, reduced limit values shall not apply before 1st January 1996.

Fig. 00.1   Outline of the new test procedure

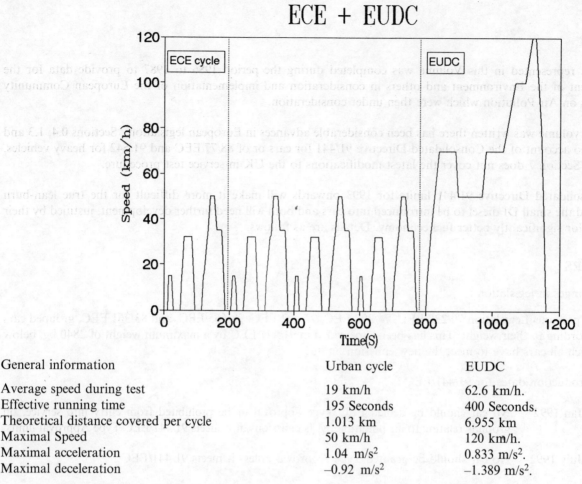

General information	Urban cycle	EUDC
Average speed during test	19 km/h	62.6 km/h.
Effective running time	195 Seconds	400 Seconds.
Theoretical distance covered per cycle	1.013 km	6.955 km
Maximal Speed	50 km/h	120 km/h.
Maximal acceleration	1.04 m/s^2	0.833 m/s^2.
Maximal deceleration	–0.92 m/s^2	–1.389 m/s^2.

- For all vehicles, check that there is not excessive smoke from the exhaust.
- For all vehicles, a general proviso that carbon monoxide emission levels will not be required to be reduced below the vehicle manufacturer's specification for the engine fitted to the vehicle.

1.1.6 Introduction of a UK in-service test procedure :

New Regulations to limit carbon monoxide and hydrocarbon emissions from in-service cars came into operation on 1st November 1991. These limits are subject to enforcement at roadside spot checks and during the annual MOT test for cars over 3 years old.

Date of car entry into service	% CO	ppm HC
1.8.83	4.5	1200
1.8.75	6.0	1200

2 HEAVY DUTY ENGINES:

2.1 Changes in Legislation

2.1.1 Section 3.4 applies to the current European Community diesel engine directive 88/77/EEC. There is however a new amending directive, 91/542/EEC, with the following limits :

	CO (g/kWh)	HC (g/kWh)	NO (g/kWh)	Pm (g/kWh) < 85kW | >85kW
Type Approval :				
(1.7.92)	4.5	1.1	8.0	0.42 | 0.36
(1.10.96)	4.0	1.1	7.0	0.15
Conformity of Production :				
(1.7.92)	4.9	1.23	9.0	0.68 | 0.4
(1.10.96)	4.0	1.1	7.0	0.15

Contents

	Page
ACKNOWLEDGEMENTS	vii
STUDY GROUP	ix
EXECUTIVE SUMMARY	xi

1. INTRODUCTION 1

    1.1 General Background 3
    1.2 Acid Emissions from the Automotive Sector 3
    1.3 The Legislative Framework 5

2. EMISSIONS CONTROL TECHNOLOGIES–SPARK IGNITION ENGINES 9

    2.1 Fuel Systems, Ignition Systems and Control Devices–Principles 11
    2.2 Current Engines 29
    2.3 Lean-Burn Engines 35
    2.4 Catalytic Systems 42
    2.5 Application of Control Technologies 50

3. EMISSIONS CONTROL TECHNOLOGIES–DIESEL ENGINES 61

    3.1 General Introduction 63
    3.2 Fuel Injection Systems and Control Devices 67
    3.3 Light Duty Engines 70
    3.4 Heavy Duty Engines 78

4. FUELS AND FUEL QUALITY 83

    4.1 Gasoline 85
    4.2 Diesel 87
    4.3 Liquid Petroleum Gas 89
    4.4 Summary 92

5. ALTERNATIVE ENGINES 93

    5.1 Stratified Charge Engines 95
    5.2 Miscellaneous 97
    5.3 Production Possibilities 108

	Page

6. MISCELLANEOUS EMITTERS — 111

   6.1 Railways — 113
   6.2 Aircraft — 113
   6.3 Diesel Power Generation — 114
   6.4 Marine Engines — 118

7. IN-SERVICE EMISSIONS PERFORMANCE AND INSPECTION — 121

   7.1 Durability of Control Systems In-Service — 123
   7.2 In-Service Inspection — 125
   7.3 Summary — 131

APPENDIX 1: BIBLIOGRAPHY — 135

APPENDIX 2: GLOSSARY — 139

APPENDIX 3: INDEX — 143

Acknowledgements

Organisations Visited
The authors of this report gratefully acknowledge the help and advice of the following organisations:

AE Developments Ltd	General Motors Corporation (USA)
Associated Octel Company Ltd	Johnson Matthey Chemicals Ltd
Austin Rover Group Ltd	Lucas Group
Citroen	Motor Vehicle Manufacturers Association of the US Inc
Commission of the European Communities	National Engineering Laboratory (DTI)
Cummins Engine Company Inc	Perkins Engines Ltd
Daimler-Benz AG	Peugeot SA
Department of Transport (UK)	Ricardo Consulting Engineers PLC
Esso Petroleum Company Ltd	Robert Bosch GmbH
Fiat Auto (including Iveco)	TNO, The Netherlands
Ford Motor Company (Europe)	Umweltbundesamt, FRG
Ford Motor Company (USA)	Volkswagen AG, FRG
Gaydon Technology Ltd	

Study Group

The following assisted in the preparation of this manual:

FELLOWSHIP OF ENGINEERING

Steering Group	:	Sir Frederick Page (Chairman)
		Mr. M. Kneale (Project Manager and Nominated Officer for the Fellowship of Engineering)

          Dr. F. Steele   Professor J.F. Davidson
          Mr. J.R. Appleton   Professor S. Eilon
          Mr. J.G. Dawson   Professor I. Fells
          Mr. G.A. Lee   Mr. R.J. Kingsley
          Professor G.F.I. Roberts Mr. V.J. Osola
          Dr. J. Gibson   Mr. K.R. Vernon
          Dr. J.H. Chesters   Dr. D. Train
          Dr. A.J. Apling   (Nominated Officer of the Department of the Environment)
          Mr. J. Murlis   (Department of the Environment)

Coal Task Group : Dr. D.R. Cope
Professor J.F. Davidson
Dr. J. Gibson (Chairman)

Oil Task Group : Mr. P. Brackley
Professor I. Fells
Mr. G.A. Lee (Chairman)
Mr. J. Solbett

Gas Task Group : Dr. C.G. James
Professor G.F.I. Roberts (Chairman)
Mr. P. Scott
Dr. F.E. Shephard
Dr. W.A. Simmonds
Professor A. Williams

Mobile Sources Group : Professor G.P. Blair
Mr. J.G. Dawson (Chairman)
Mr. A. Silverleaf

FOSTER WHEELER POWER
PRODUCTS : Dr. R. Fletcher
Mr. K. Johnson
Mr. H. Luaw
Mr. D. McSherry
Mr. H.T. Wilson (Programme Manager)

HOY ASSOCIATES (UK) : Mr. D.W. Gill
Mr. H.R. Hoy (Director)
Mr. A.G. Roberts
Mr. J.E. Stantan
Mr. D.M. Wilkins

WARREN SPRING
LABORATORIES : Dr. M. Williams
Mr. J. Potter
Dr. J.H. Weaving

Executive Summary

0.1 Background

This handbook forms part of a wider study on the acidic emissions from a range of combustion technologies. The study was commissioned with The Fellowship of Engineering by the Department of the Environment and deals with the combustion of solid, gas and petroleum fuels in stationary sources and, in this handbook, with mobile sources. The objective of the study is to provide information on the emissions, efficiencies and costs of control and abatement technologies currently in use and which are likely to be used in the next 15 years or so.

0.2 Content

This handbook on mobile sources is chiefly concerned with motor vehicles, both gasoline and diesel engined. Other sources are considered briefly but these contribute only a small amount to acidic emissions. This handbook will deal primarily with the so-called 'regulated' pollutants from motor vehicles, namely oxides of nitrogen (NO_x), carbon monoxide (CO) and total hydrocarbons (THC or often simply HC); the term volatile organic compounds (VOC) is increasingly used. Although in terms of overall acid deposition, it is generally considered that sulphur dioxide (SO_2) is the single most important pollutant, motor vehicles are not significant sources of SO_2 on a national or international scale. Emissions of NO_x however can contribute significantly to the acidity of deposition via one of their important oxidation products in the atmosphere, nitric acid (HNO_3). Hydrocarbon emissions while not acidic in themselves contain components which may give rise to concern over health effects. The hydrocarbons emitted from vehicles and other sources also contribute to the production of acidic end products in the atmospheric chemistry processes. They generate reactive radicals and other species such as ozone and hydrogen peroxide which in the photochemical cycle can eventually lead to the oxidation of SO_2 and NO_x in the gas phase and in rain and cloud water.

The study deals with a range of motor vehicle technologies. For gasoline engines, conventional current technologies are discussed and form a baseline from which improved pollution control technologies and their costs are evaluated. These include descriptions of fuel, engine management and ignition systems. Future developments in lean burn engines and catalysis systems are also discussed. Consideration is given to the technologies, their emission performance and where available, their associated costs.

Diesel engines are also considered and current and developing technologies for both light duty and heavy duty diesels are described. Current and future developments in gasoline and diesel fuel quality is given and alternative engine technologies are also discussed. Finally a brief section on mobile sources other than road transport is presented, including aircraft, railways, etc.

0.3 Motor Vehicle Emissions in the Context of National Acidic Emissions

It has already been noted that while motor vehicles in most developed Western countries are a minor source of SO_2, they can be a significant source of NO_x. These issues are addressed in the report in the context of UK emissions, and in this summary some brief conclusions only are appropriate. Summaries of national UK emissions of SO_2, NO_x, HC (VOC) and CO are shown in Figures 1–4.

For the UK in 1985, of the estimated total emission of 3.58 million tonnes of SO_2 some 71% arose from power stations, with industrial sources contributing a further 15%. Road transport accounted for only 1% of the total. The position for NO_x was quite different, with road transport and power stations here both accounting for 40% of the total emission of 1.84 million tonnes, with industrial sources accounting for 10%. For hydrocarbons, or volatile organic compounds, the main source is not fuel combustion but evaporative losses, either from solvent use or through uncontrolled leakage or losses. In the motor vehicle these can arise from the carburettor as 'hot-soak' losses, from the fuel tank, particularly during warmer weather, or from tank filling. Road transport in the UK in 1985 accounted for 26% of the national total emission of hydrocarbons of 2.06 million tonnes, while emissions from industrial processes and solvent use amounted to 44%. Natural gas leakage amounted to 20% but this is largely composed of methane which is very unreactive in the atmosphere. Of the road transport contribution 22% arose from petrol evaporation, the rest coming from exhaust emissions of unburned and partially burned fuel components.

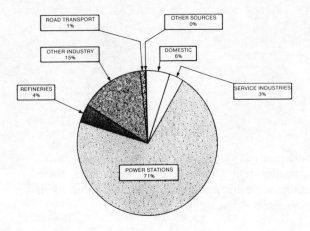

Figure 0.1 UK Sulphur Emissions 1985–Total = 3.58 Million Tonnes SO_2

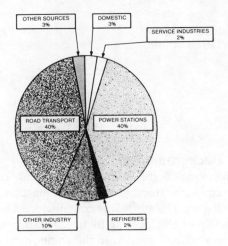

Figure 0.2 UK Nitrogen Oxides Emissions 1985–Total = 1.84 Million Tonnes NO_2

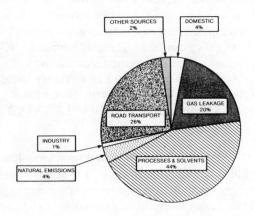

Figure 0.3 UK Volatile Organic Compound Emissions 1985–Total = 2.06 Million Tonnes

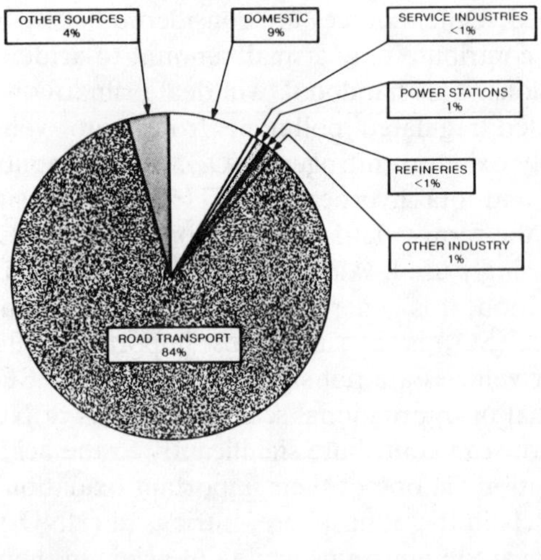

Figure 0.4 UK Carbon Monoxide Emissions 1985–Total = 5.39 Million Tonnes

Road transport is the dominant source of UK emissions of CO amounting to some 84% of 1985 emissions, which totalled some 5.39 million tonnes.

0.4 The Legislative Position

Since 1970 gasoline engined vehicle emissions in the European Economic Community (EEC) have been regulated by a series of Directives, shown in Table 0.1, which have been based on vehicle weight and have become increasingly stringent. The series of Directives are based on the Economic Commission for Europe (ECE) Regulation 15 and its subsequent amendments, the most recent of which is 15-04; they are all referred to a standard test cycle of average speed 18.7 km/h which was originally designed to simulate a typical vehicle drive in urban conditions but run on a chassis dynamometer under carefully controlled laboratory conditions.

Successive amendments to the original Regulation 15-00 have become increasingly stringent. Looking at cars of reference weight 850–1020 kg, the emission limit for CO has decreased by 50% from a value of 117 g/test in the original ECE Regulation 15, to a value of 58 g/test in Amendment 04. It is less straightforward to compare the individual HC and

NO$_x$ limits in 15-00 and 15-04 as only a combined HC and NO$_x$ limit is quoted in 15-04, and no NO$_x$ limit was quoted in 15-00. From 15-00 to 15-03, the HC limit decreased by 25% from 8.7 g/test to 6.5 g/test. The HC+NO$_x$ limit decreased from 15-02 to 15-04 from 27.0 g/test to 19.0 g/test, a decrease of 30%.

In 1985, after considerable discussion, most of the Member States of the EEC agreed a new series of emission limits which have become known as the 'Luxembourg Agreement' and which are shown in Table 0.2. These limits are defined by engine size rather than vehicle weight. Moreover, Member States are also discussing extensions to the test cycle which will incorporate sections of higher speed driving.

In terms of the limit values given in Tables 0.1 and 0.2, the 'Luxembourg Agreement' represents a substantial decrease over Amendment 15-04. Comparisons with the Luxembourg Agreement are not straightforward given that 15-04 is framed in terms of vehicle weight and the Luxembourg limits are given in terms of engine size. However to a good approximation it can be assumed, for the UK at least, that an average of the 850–1020 kg and 1020–1250 kg limits gives the 'UK fleet average'

limit. Similarly using figures of 8.3%, 38.3% and 52% for the proportions of 2 litre, 1.4–2 litre and 1.4 litre cars the fleet-average 'Luxembourg Limits' can be derived. (These figures are SMMT data for 1985 published by the Society of Motor Manufacturers and Traders; 1.3% of vehicles are classified as 'unknown'.) On this basis the fleet average CO limits are 62.5 g/test for 15-04 and 37.0 g/test for Luxembourg a decrease of 41%. For HC+NO$_x$ the corresponding figures are 19.8 g/test and 11.4 g/test, a decrease of 42%.

The new limits contained in Table 0.2 consequently impose significant demands on vehicle manufacturers and a great deal of engine and vehicle development is linked to the achievement of these more stringent emission limits. The discussion of future technological developments in road transport given in the handbook is therefore very closely linked with future legislation.

The Luxembourg limits for small cars are less stringent than those for the medium or large category, although Stage II limits for small cars are still to be agreed. It is possible therefore that, given the fraction of small cars in the UK fleet (52% in 1985), the greater part of vehicle emissions of CO and HC+NO$_x$ could arise from this category of

Table 0.1 ECE Regulation 15: Type Approval Standards

Reference Weight (kg)	Original ECE 15-00 EEC Directive 70/220 Date: From 1971		ECE 15-01 Amendment EEC Directive 74/290 Date: October 1975 (ECE)		ECE 15-02 Amendment EEC Directive 77/102 Date: March, 1977 (ECE)			ECE 15-03 Amendment EEC Directive 78/665 Date: October, 1979 (ECE)			ECE 15-04 Amendment† EEC Directive 83/361 Date:	
	g test^{-1}		g test^{-1}		g test^{-1}			g test^{-1}			g test^{-1}	
	CO	HC	CO	HC	CO	HC	NO$_x$	CO	HC	NO$_x$	CO	HC + NO$_x$
TYPE 1 test												
750	100	8.0	80	6.8	80	6.8	10	65	6.0	8.5	–	–
750–850	109	8.4	87	7.1	87	7.1	10	71	6.3	8.5	–	–
850–1020	117	8.7	94	7.4	94	7.4	10	76	6.5	8.5	58	19.0 (1020 kg)
1020–1250	134	9.4	107	8.0	107	8.0	12	87	7.1	10.2	67	20.5
1250–1470	152	10.1	122	8.6	122	8.6	14	99	7.6	11.9	76	22.0
1470–1700	169	10.8	135	9.2	135	9.2	14.5	110	8.1	12.3	87	23.5
1700–1930	186	11.4	149	9.7	149	9.7	15.0	121	8.6	12.8	93	25.0
1930–2150	203	12.1	162	10.3	162	10.3	15.5	131	9.2	13.2	101	26.5
2150	220	12.8	176	10.9	176	10.9	16.0	143	9.6	13.6	110	28.0

†In 15-04 HC measurements are made using FID instruments. Earlier Amendments used NDIR instruments. An approximate conversion to FID values is obtained by multiplying NDIR values by 2.3.

Table 0.2 The 'Luxembourg Agreement' for Light Duty Vehicles

Engine Size (litres)	Emission Limit (g/test)				Introduction Dates	
	CO	HC	NO$_x$	HC + NO$_x$	New Models	All
2.0	25.0	3.0	3.5	6.5	Oct 1990	Oct 1991
1.4–2.0	30.0	–	–	8.0	Oct 1991	Oct 1993
1.4	45.0	9.0	6.0	15.0*	Oct 1988	Oct 1989

*Stage I limit, Stage II to be agreed.

vehicle. Some caution is necessary however since emission limits may not reflect accurately the actual emissions in service. Measurements will be necessary in the future to address this point.

In terms of diesel engine emission legislation, the ECE Regulation 49 applying to heavy duty diesels was adopted in 1982. However this has not been adopted in national or EEC homologation procedures. The Commission of the European Communities (CEC) has proposed reductions on Regulation 49 of 20% for CO and NO_x and 30% for HC, on the current Regulation 49 emission limits of 14 g/kWh, 18 g/kWh and 3.5 g/kWh for CO, NO_x and HC respectively. Particulate emission limits for heavy duty diesels are not at present under formal discussion.

Light duty diesels are covered by the Luxembourg limits for CO, HC and NO_x, and emission limits for particulates are under discussion within the EEC.

The implications of these regulations for diesel engine technologies are discussed in Section 0.10 below.

0.5 Control Devices, including Fuel and Ignition Systems

The pollutants from gasoline engines are shown in graphical form in Figure 0.5. It will be seen that the magnitude is greatly influenced by the air/fuel (A/F) ratio of the mixture that is drawn into the engine and subsequently burned in the combustion chamber. The dotted line indicates the stoichiometric ratio of air to fuel which is the exact chemical quantity of air required to burn the fuel completely. Unfortunately due to the difficulty of perfect mixing in the very short time available, combustion is not quite 100%. On the left of the dotted line there is excess fuel and insufficient oxygen from the air (79% nitrogen + 21% oxygen) and hence HC and carbon monoxide are relatively high; for leaner (i.e. less fuel) settings both these pollutants are reduced but not eliminated. NO which is formed in the combustion chamber from the nitrogen and oxygen in the air due to the high temperature of combustion reaches a maximum just weak of the stoichiometric A/F ratio. At richer (i.e. more fuel) A/F ratios there is insufficient oxygen, and as settings are leaned off the temperature of combustion is reduced and less nitrogen and oxygen combine. As the nitric oxide formation increases with the temperature of combustion the control of the ignition point by the spark is critical. Retardation reduces the temperature and hence the NO but unfortunately also reduces the efficiency of combustion with consequent deterioration of fuel consumption. On the right of the dotted line (Figure 0.5) it can be seen that by weakening the mixture to A/F ratios beyond about 16:1 substantial reductions in NO formation are achieved.

It will be appreciated from the above discussion that the accurate control of air/fuel ratio and spark timing are vital for the reduction of pollutants.

Exhaust gas recirculation (EGR) may be used to reduce NO formation in the combustion chamber. This is accomplished by feeding up to 10% of the exhaust gas back into the induction system which reduces the temperature of combustion. It is an inexpensive method but is not universally favoured because of possible contamination of the intake system.

In the document the basic principles and descriptions of current and developing fuel systems are described, including carburettors, throttle body (single point) and multi-point fuel injection systems. The operating principles of conventional ignition systems are described as are transistorised and electronic systems incorporating electronic control units. This discussion is followed by a section describing engine management systems integrating ignition and fuel systems.

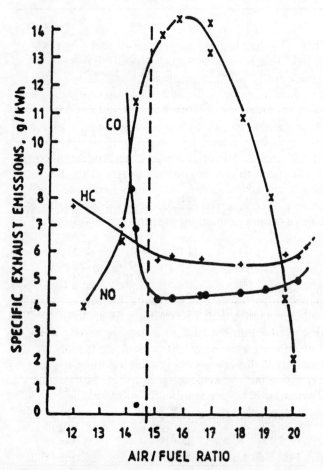

Figure 0.5 Variation of Emissions with Air/Fuel Ratio

0.6 Current Engines

In the context of this handbook, the term 'current' is used to mean those technologies employed to meet the European Regulation 15 up to and including Amendment 04. A wide range of variants on basic engine technologies are in existence for a range of commercial marketing and technical reasons. In planning to comply with the European regulations up to and including 15-04, manufacturers have in general opted for improvements to engine efficiency, tighter tolerances, improvements to fuel and ignition systems, and an adoption of engine management systems rather than the introduction of more fundamentally different technologies such as full lean burn engines or catalyst systems.

0.7 Lean Burn Engines

The basic concept of the high-compression lean burn (HCLB) engine is to attempt to run the gasoline engine at A/F ratios of 18 to 20:1 or leaner at which ratios it will be seen from Figure 0.5 that all three pollutants are low. However this is not easy to accomplish and needs much fundamental research in combustion chamber design. (It should be noted that Figure 0.5 shows measurements on one particular engine and should be regarded as illustrative of the broad pattern of pollutant behaviour only.)

The HCLB engine has a basic advantage in giving enhanced fuel economy as pumping losses are reduced in a similar manner to the diesel engine. Current production 'lean burn engines' on sale in the UK run at about 17–18/1.

The advantages of the lean-burn approach to emission reduction, compared with catalyst methods for example, are the improved fuel efficiency, and the fact that the HCLB engine is an engineered solution with the consequent potential for inherent robustness. The fuel economy advantage over rival systems such as three-way catalyst equipped vehicles (which are constrained to run at stoichiometric–see below) is not easy to estimate as direct comparisons are difficult but it is broadly accepted that an HCLB vehicle with exhaust gas recirculation would show an advantage of 5–10% over a three-way catalyst equipped car.

One of the disadvantages of the HCLB engine is the reduced driveability of the vehicle, particularly under the transient conditions experienced on the road where engine mis-fire can occur. This places great demands on the fuel and ignition control systems which must be taken into account when comparing the costs of HCLB vehicles with those employing other technologies. Furthermore, it is likely (see below) that, in order to meet the so-called 'Luxembourg' limits, vehicles with 1.4–2 litre HCLB engines will also need oxidation catalysts to reduce the hydrocarbon emissions sufficiently, thus increasing the production costs.

0.8 Three-Way Catalyst Systems

Three-way catalysts are so called because they control CO and HC emissions by oxidation (to CO_2 and water) and NO_x by reduction in the presence of CO and H_2 to nitrogen, CO_2 and water. In order to control all three pollutants efficiently, three way catalysts (TWCs) must operate in a relatively narrow window of A/F ratios around stoichiometric. If this can be achieved then TWCs can produce large reductions in all three regulated pollutant emissions. It can be concluded that TWC vehicles would be capable of meeting the so-called Luxembourg limits with closed loop control and with improved engine management systems. The closed loop control involves sensing the A/F ratio from the exhaust gases as they leave the engine by means of an oxygen sensor and controlling the A/F ratio at stoichiometric with an electronic control unit (ECU).

The disadvantages with TWC systems are first the reduced fuel economy which arises from the need to operate at stoichiometric A/F ratio and at considerable extra cost compared with an HCLB system. However it is worth noting that with the improved engine management systems, A/F control etc. can achieve fuel economies similar to those of a 'baseline' vehicle with a spark ignition engine and carburettor (and no catalyst). Secondly, another problem with TWC cars is misfuelling, i.e. the use of petrol containing lead in place of unleaded fuel which is required to avoid poisoning the catalyst. There are steps which can be taken to ameliorate this problem although it would be difficult to overcome completely.

0.9 Technologies, Legislation and Costs

It has already been noted that future technological development will be encouraged by the Luxembourg emission limits in the three engine size categories, less than 1.4 litres, 1.4–2.0 litres and more than 2.0 litres. Approximate production costs are given here and are based on costs additional to current 15-04 vehicles, at 1987 prices. The costs are given as production costs. It is difficult to give accurate estimates of the final cost to the consumer as these can be influenced by a variety of commercial considerations. However as a rule of thumb, a multiplier factor of 2.5–3 is often used.

In the smaller category (1.4 litres) it is almost the unanimous view of the motor industry that a lean burn approach will meet the limits, but this conclusion is dependent on the more stringent levels to be discussed within the EEC as a stage II beyond the Luxembourg agreement. At present the use of catalysts in this category is considered to be too expensive but the evidence is equivocal on this issue. Two alternatives are feasible for most vehicles:

(i) Moderately lean-burn (A/F c. 18:1) + EGR, production cost £30.

(ii) Full lean-burn (improved fuel consumption) plus feedback control, production cost c. £200.

In the middle size range, 1.4–2.0 litres, both TWC and HCLB plus oxidation catalysts are options, but most manufacturers appear to favour the latter. The additional costs for lean burn plus oxidation catalyst would be £200. For TWC systems with full engine management and single point (throttle body) injection the costs would be £330–400.

For the large car category, TWC is at the present time the only practicable way of achieving the Luxembourg limits, as no HCLB engines of this size are near production. According to the degree of sophistication of the fuel system, etc., a TWC system for large cars would cost £400–£550.

A summary of the emissions, fuel economies and costs for the more important gasoline engine technologies is shown in Table 0.3.

0.10 Diesels

Diesel engines in many ways offer technical advantages over gasoline engines. In the context of this handbook on acid emissions the most important of these are that the diesel engine could be described as the ultimate 'lean-burn' engine in that it operates at air/fuel ratios of 20–40:1 or even higher. Diesels are consequently low polluting vehicles as far as the gaseous pollutants CO, HC and NO_x are concerned. Direct comparisons are difficult but compared with a broadly equivalent gasoline engined vehicle, CO and HC emissions from diesels can be up to an order of magnitude lower in urban driving conditions with NO_x emissions up to roughly a factor of 2–3 lower. Fuel economy advantages also exist and can be up to 30–50% better than the broadly equivalent gasoline vehicles in urban driving conditions. However at higher power loadings and speeds these differences can be greatly reduced.

Particulate emissions arising from incomplete combustion can be a problem for diesels compared with gasoline engined vehicles especially when the vehicle is poorly maintained. Diesel fuel also contains roughly ten times as much sulphur as gasoline so that per unit mass of fuel consumed, SO_2 emissions from diesels are much greater than from gasoline vehicles. While this may in some specific locations be an issue, in terms of national and international emissions and the acid deposition debate, the diesel contribution has already been shown to be very small.

Current diesel engine technologies are conveniently divided into two groups differentiated by the type of combustion chamber. Direct Injection (DI) diesel engines are the predominant type in the heavy duty vehicle sector, and here, as the name implies, the fuel is injected directly into a single combustion chamber formed by the cylinder head and piston. The light duty diesel sector employs mostly Indirect Injection (IDI) where the fuel is injected into a high swirl pre-chamber containing a glowplug which is used to heat the fuel on cold starting before it passes into the main combustion chamber over the piston. However there are some recent developments of DI engines in the light duty sector, primarily for the potentially lower fuel consumption.

In terms of technologies and legislation, an existing EEC Directive (embodying the ECE Regulation 49) applying to gaseous emissions of NO_x, CO and HC from Heavy Duty diesels has not been ratified by Member States. Discussions are currently under way to define more stringent emission limits for adoption within the Community. The precise values of the limits have yet to be agreed but it is likely that heavy duty diesel DI engines would need relatively sophisticated electronic controls on fuel injection quantities and timing in order to meet the limits being discussed. Particulate emissions can be of more concern in the context of diesel engines than gasoline engines, but European legislation on particulate emissions from heavy duty diesels has not been formally proposed at this stage.

Table 0.3 Emissions, Fuel Economy and Costs for Some Vehicle Technologies (on a tuned, new vehicle basis)

Technology	ECE R15 Cold Start Emission (g/test)				Fuel Economy Index	Production Cost (£ 1987)
	CO	HC	NO_x	HC+NO_x		
Luxembourg Limits small†	45	–	6	15	–	–
medium	30	–	–	8	–	–
large	25	3.0	3.5	6.5	–	–
Baseline*	70	15	8.0	23.0	–5%	0
15-04 Current‡	50–60	9.12	6–8	15–20	0	90–120
Luxembourg HCLB + Ox Cat. + sensor + TBI	15	2–3	3–5	5–8	+5–10%	300–400
TWC + full e.m.s. + TBI	15	2–3	2	4–5	–5%	500–600

*Conventional spark ignition, carburettor, contact breaker ignition
†Stage I limits; Stage II limits for small cars still to be discussed
‡Breakerless ignition, improved carburation, automatic choke

Light duty diesel vehicles (cars, etc.) are covered by the 'Luxembourg' limits for the gaseous pollutants CO, HC and NO_x. Diesel engined vehicles of less than 1.4 litres are required to meet the same standards as gasoline engines but *all* light-duty diesel cars of engine capacity greater than 1.4 litres are required to meet the gaseous standards for gasoline engines in the 1.4 to 2.0 litre range. The low polluting characteristic of diesels has already been noted and the current IDI technologies in use by light duty vehicles should prove adequate to meet the limits. Although not covered by the Luxembourg agreement, particulate emission standards for light duty diesels are being discussed. These are likely to be gravimetrically based and it seems probable that light duty vehicles will be able to meet the limits with current IDI technology. Developments of light duty DI engines will improve fuel economy, probably by up to 10%, but there will be a trade-off between decreased particulate and increased NO_x emissions. The precise implementation of the DI light duty diesels will therefore be governed primarily by the balance between the limits for particulate and NO_x emissions eventually agreed within Europe.

One aspect of the costs of diesel vehicles compared with other technologies has already been covered in the foregoing discussion of fuel economy of the diesel engine; however the costs to the customer will of course be determined by the particular fuel excise and pricing policies in existence in various countries. Production costs of diesel engines are overall currently higher than gasoline engines and indeed point-of-sale diesel vehicle costs can be significantly higher than for gasoline vehicles. Overall in the current commercial climate regarding fuel and vehicle pricing, diesel vehicles show economic advantages to the customer for mileages not significantly greater than average.

0.11 Fuel Quality
In terms of future developments in gasoline quality the most important feature will be the increasing availability and use of low lead (0.15 g/litre) and unleaded petrol (0.013 g/litre). EEC Directive 85/210/EEC requires Member States to make unleaded petrol widely available by October 1989 and some countries have already done this. The UK plans to achieve wide availability of unleaded petrol by 1988. The rate at which unleaded petrol will be taken up in Europe will be determined not only by the rate of phasing out of vehicles which are not capable of running on unleaded petrol, but also on the fiscal policies employed in different countries for leaded and unleaded fuels.

A most important implication of the use of unleaded petrol for vehicle manufacturers and technologies is that the lead additives as well as reducing knock also act as lubricants for valve seats and stems. Removal of the lead additives means that manufacturers will need to harden the valve seats. On some vehicles this is already done. Unleaded petrol will have a rating of 95 Octane Number, compared with current premium grades of 98, so that there will be some loss in efficiency and fuel economy. This is likely to be of the order of 5% and may even be offset by the use of more sophisticated engine management systems.

In order to maintain the octane number at 95 in unleaded petrol other compounds may have to be added. Most important among these are the aromatic hydrocarbons and the aforementioned Directive sets a limit of 5% on the benzene content of unleaded petrol. However it is possible that the total aromatic content of gasoline will increase which will have an effect on the potential of gasoline emissions for the formation of photochemical ozone. At this stage it is not possible to quantify this effect. It is also possible that an increased aromatic content and associated changes in fuel volatility might produce vapour lock effects in some vehicles.

Changes in diesel fuel quality are also likely to take place, primarily as a result of an increased demand on the refining process to produce more diesel fuel. It is likely that fuel quality (as measured by Cetane Number) will decrease with the main effect being to increase particulate emissions (and also to reduce NO_x emissions to some extent); it is also likely that another effect of a decrease in cetane numbers would be to make starting more difficult.

0.12 Alternative Engines
The development of engines other than current spark ignition and diesel engines is discussed. The discussion covers two stroke engines including two stroke diesels, rotary engines, gas turbines, Stirling cycle and Rankine cycle engines. While many of the developments are of technological interest, apart from the stratified charge engine developed by Honda, and some two stroke engines in earlier years, there has been no significant production of alternative engines. This is probably a reflection of the often considerably greater complexity and expense coupled with the fact that emission regulations have so far been met with more conventional engines. It is difficult to make definitive projections of future developments in alternative engines in the mobile source sector. However in some areas developments are under way which allow some tentative statements to be made. In the gas turbine field, for example, component development programmes, particularly in ceramics, are likely to result both in improved efficiency and reduced cost. It is possible that gas turbines in the 300–500 HP range will be used in heavy trucks towards the end of the century. The most promising approach to the further development of the gasoline spark ignition engine appears to be in the stratified charge concept already noted above.

Developments are under way in adiabatic and compound engines and it is possible that increased investment in this development may take place in future.

0.13 Miscellaneous Emitters

Emissions from other mobile sources such as railways, aircraft, ships, etc, are small on a national scale compared with those from road transport and stationary sources (see Figures 0.1 to 0.4). Moreover in general there is very much less information available on such sources in the European context. Where data are available they predominantly originate in the USA and any application in the UK or the rest of Europe should only be carried out bearing the likely uncertainties fully in mind. Although on a national basis, the emissions from sources considered in this section are small, they may in certain circumstances give rise to problems locally.

0.14 In-Service Inspection

A brief discussion is given of in-service inspection as a procedure designed to ensure that emissions standards are maintained in practice. Procedures and limits for petrol-engined vehicles in use in various countries are summarised, namely Austria, Canada, France, FRG, Japan, Netherlands, Sweden, Switzerland and the USA. A limit on idle CO emissions is common to all countries and a value of 4.5% is used in all except Austria and the FRG where a value of 3.5% is used. (Earlier in the FRG a limit of 4.5% was used.) In the USA a limit of 1.2% CO is applied to catalyst cars. Some countries (Austria, Canada, Japan and the USA) also set limits on hydrocarbons at idle. The limits vary from 220 ppm in the USA (for catalyst cars) to 1200 ppm in Japan and Canada, with 600 ppm being used in Austria.

Some in-service measurements are discussed, notably work in the FRG which showed that of a recent sample of 2000 cars equipped with conventional ignition systems, only 19% were found to be in accord with manufacturers' specifications. In the case of 500 cars fitted with electronic ignition systems, 48% were found to be correctly adjusted.

The instrumentation required for in-service inspection in terms of gas analysers and other capital equipment is described, in-service inspection of diesel vehicles is discussed briefly.

1. Introduction

1.1 General Background

1.2 Acid Emissions from the Automotive Sector

1.3 The Legislative Framework

1. Introduction

1.1 General Background

This handbook on the acidic emissions from mobile sources has been compiled by the Fellowship of Engineering in response to a commission from the Department of the Environment (DoE) to study control and abatement technologies as applied in the automotive field. The most important groups of engines are the petrol or spark ignition (SI) and compression ignition (CI), i.e. diesel engine types. Whereas road transport predominates in importance with regard to total emissions to the atmosphere of acid gaseous species and photochemical precursors from mobile sources, attention has also been directed to diesel power generation, aviation, motorcycles and marine craft.

Where possible emission data for the pollutants of interest over the full performance range of the vehicle have been sought. In practice only a few organisations in Europe have been able to supply such data. These organisations have made such investigations for national government purposes, whereas vehicle manufacturers would normally test vehicles in accord with the prescriptions of the legislation appropriate to the particular market place.

The data base sought was world-wide and a list of organisations visited is given in the acknowledgements. These organisations may be broadly classified as:

(i)   industrial;
(ii) research (industry and government); and
(iii) government (UK, FRG and EC).

Prior to a visit an agenda in the form of a questionnaire was despatched. This agenda was modified according to the particular organisation's sphere of activity but had a common format.

BACKGROUND AND APPROACHES TO EMISSION REDUCTION

Although the remit of this inquiry was for acidic pollutants, because of existing and proposed legislation the range of regulated pollutants must be taken into consideration.

Passenger motor vehicles in Europe are largely propelled by spark ignition engines, though there is a growing usage of diesel engines, particularly in Germany and in countries where tax incentives help to make them an economic alternative. Commercial vehicles are almost universally fitted with diesel engines in Europe. It is difficult to predict the proportion of diesel to spark ignition engines in future years as this is much influenced by fiscal and political considerations.

The principal pollutant gases produced in the combustion chambers of spark ignition engines and indeed diesel engines are carbon monoxide (CO), unburnt hydrocarbons (HC) and nitric oxide (NO). Diesel engines emit a small amount of nitrogen dioxide (NO_2) under certain engine loan conditions, and also a little sulphur dioxide (SO_2) derived entirely from the sulphur content of the fuel.

These so-called primary pollutants are of interest in their own right, but these can also give rise to chemical reactions in the atmosphere when hydrocarbons and nitric oxide together with the oxygen and water vapour in the air can undergo complex reactions forming nitrogen dioxide (NO_2), nitric acid (HNO_3) and ozone (O_3) as secondary products. Moreover, the processes forming these particular secondary pollutants are also important, for example, in oxidising SO_2 (primarily emitted by non-vehicular sources – see Section 1.4) to sulphates which play a significant role in acid deposition. Collectively the various oxides of nitrogen are known as NO_x and are referred to as such in vehicle legislation.

CO, HC and NO_x pollutants have been considerably reduced to meet EEC regulations and can be reduced still further but not without additional expense. While carbon monoxide and hydrocarbons can be reduced without deterioration in fuel consumption, it is very difficult to reduce oxides of nitrogen without impairing the engine efficiency.

1.2 Acid Emissions from the Automotive Sector

In the United Kingdom (UK) mobile sources are estimated to contribute about 40% of anthropogenic nitrogen oxides emissions. Current best estimates suggest that this 40% is made up of 11% from diesel

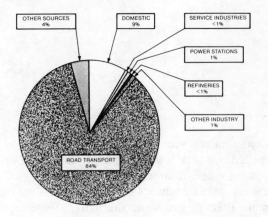

Figure 1.1 UK Carbon Monoxide Emissions 1985–Total = 5.39 Million Tonnes

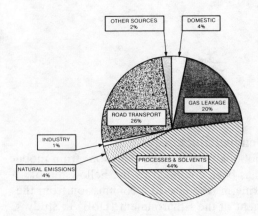

Figure 1.2 UK Volatile Organic Compound Emissions 1985–Total = 2.06 Million Tonnes

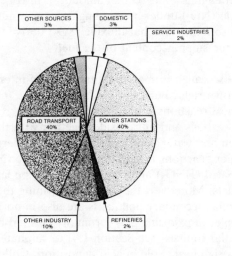

Figure 1.3 UK Nitrogen Oxides Emissions 1985–Total = 1.84 Million Tonnes NO_2

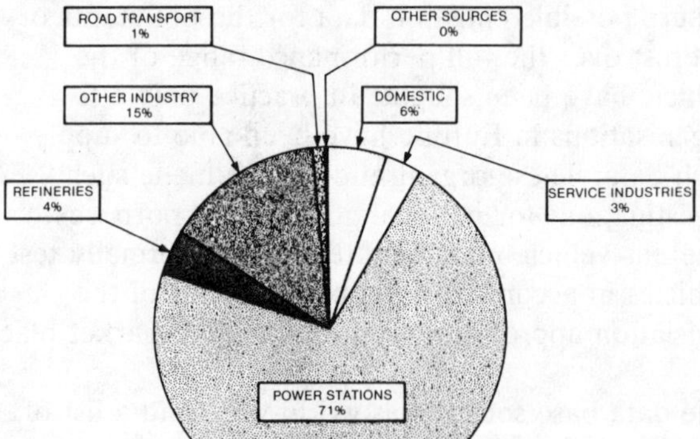

Figure 1.4 UK Sulphur Emissions 1985–Total = 3.58 Million Tonnes SO_2

vehicles and 29% from gasoline vehicles. Heavy duty diesel engined vehicles emit considerably more NO_x per unit distance driven than light-duty gasoline vehicles. For example, light-duty passenger cars may emit about 1–2 g km^{-1} but heavy-duty trucks emit approximately 30 g km^{-1} of nitrogen oxides.

As increasingly stringent standards for NO_x emissions from motor vehicles of the light-duty M1 class are introduced, then NO_x emissions from heavy-duty diesel engines will take on an increasing importance.

Sulphur dioxide emissions from gasoline engines are negligible (about 0.3% of total SO_2 emissions from all sources). However, SO_2 emissions from diesel engines in the UK have been estimated to be approximately 41 kilotonnes which is about 3–4 times the total emission from gasoline engines.

The relative contributions of mobile and stationary sources to total UK emissions are given in Figures 1.1, 1.2, 1.3 and 1.4. Figure 1.5 shows the distribution of vehicle emissions according to road class. For light-duty petrol engined vehicles Figure 1.6 shows the distribution of engine size in the UK fleet. Thus, 52.1% of vehicles are in the less than 1.4 litre range and 38.3% are in the medium engine size range of 1.4 to 2.0 litres.

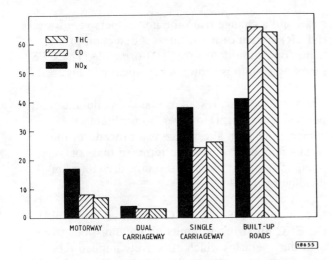

Figure 1.5 % Contribution to Total Petrol Engined Vehicle Emissions of 4 Road Classes

1.3 The Legislative Framework

THE EUROPEAN SCENE

Historically, the regulations concerning the type approval (TA) of new light duty vehicles equipped with SI and CI engines have been negotiated under the auspices of the United Nations Economic Commission for Europe (ECE). After considerable technical and legal discussion in Europe with respect to consideration of possible barriers to trade, the Economic Commission for Europe Regulation 15 (ECE R15) 1970 was adopted by the Commission of the European Communities (CEC) and published as a Community Directive (70/220/EEC). Initially ECE R15 applied to carbon monoxide (CO) and hydrocarbons (HC) only; however, in 1977 the regulation was amended to include nitrogen oxides (NO_x). This amendment was known as ECE R15-02. The third amendment to the regulation (ECE R15-03) became effective in October 1979 and resulted in a reduction of CO, HC and NO_x limits by approximately 19, 11 and 15% respectively. The various Type Approval standards of ECE R15 up to the fourth amendment are given in Table 1.1.

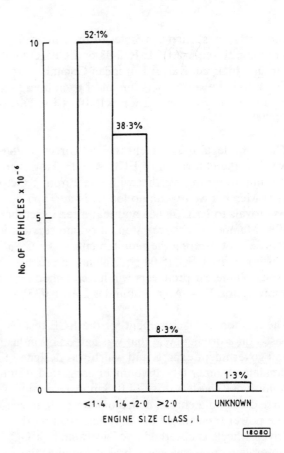

Figure 1.6 Distribution of Petrol Engined Vehicles in the UK by Engine Size

Table 1.1 ECE Regulation 15: Type Approval Standards

Reference Weight (kg)	Original ECE 15 EEC Directive 70/220 Date: From 1971		ECE 15-01 Amendment EEC Directive 74/290 Date: October 1975 (ECE)		ECE 15-02 Amendment EEC Directive 77/102 Date: March, 1977 (ECE)			ECE 15-03 Amendment EEC Directive 78/665 Date: October, 1979 (ECE)			ECE 15-04 Amendment EEC Directive 83/351 Date:	
	g test^{-1}		g test^{-1}		g test^{-1}			g test^{-1}			g test^{-1}	
	CO	HC	CO	HC	CO	HC	NO_x	CO	HC	NO_x	CO	HC + NO_x
TYPE 1 test												
750	100	8.0	80	6.8	80	6.8	10	65	6.0	8.5	–	–
750–850	109	8.4	87	7.1	87	7.1	10	71	6.3	8.5	–	–
850–1020	117	8.7	94	7.4	94	7.4	10	76	6.5	8.5	58	19.0 (1020 kg)
1020–1250	134	9.4	107	8.0	107	8.0	12	87	7.1	10.2	67	20.5
1250–1470	152	10.1	122	8.6	122	8.6	14	99	7.6	11.9	76	22.0
1470–1700	169	10.8	135	9.2	135	9.2	14.5	110	8.1	12.3	87	23.5
1700–1930	186	11.4	149	9.7	149	9.7	15.0	121	8.6	12.8	93	25.0
1930–2150	203	12.1	162	10.3	162	10.3	15.5	131	9.2	13.2	101	26.5
2150	220	12.8	176	10.9	176	10.9	16.0	143	9.6	13.6	110	28.0

Currently, most Member States of the European Community require ECE R15-03 or ECE R15-04 through their equivalent European Commission Directives; however, Belgium and Luxembourg have mandated only as far as ECE R15-01 (EEC Directive 74/290).

The UK's legal requirement is EEC Directive 78/665 which is the equivalent of ECE R15-03. However, it is important to note that with effect from 1 October 1984 Member States can no longer issue Type Approvals to EEC 78/665 and that after 1 October 1986 Member States *may* stop entry into service of vehicles not meeting the requirements of EEC 83/351 which is the ECE R15-04 equivalent. In practice most volume car producers will have applied or are applying for Type Approval under EEC 83/351.

The method of measurement for the ECE tests is based on a driving cycle that was agreed by industry and government experts and which was designed to simulate an inner city drive under congested traffic conditions. The test drive (4.05 km in 13 minutes) is carried out on a chassis dynamometer (rolling road). The power transmitted to the rolling road by the driven wheels is absorbed by, for example, a DC motor/generator or a water brake. Initially, the exhaust gas handling system was relatively simple, all the exhaust gases discharged being collected in a large plastic bag. At the end of the test the concentrations of the pollutants of interest were analysed and the mass emissions of the gases calculated. The fourth amendment to the regulation (ECE R15-04) introduced the constant volume sampling system (CVS) for the handling of the exhaust gases and a change from the non-dispersive infrared (NDIR) system of hydrocarbons detection to the flame ionization detector (FID) principle. The CVS sampling system is shown schematically in Figure 1.7.

The results are expressed as mass of pollutant emitted per test (g test^{-1}). (However, to facilitate comparisons with alternative test procedures the results may be expressed in terms of mass of pollutant emitted per unit distance driven (g km^{-1}) dividing by 4.052.)

In November 1983 the FRG submitted proposals to the EEC for stringent limits on the emissions from light-duty vehicles which effectively implied the application of catalyst technology to clean up emissions. These standards also would have required the US Federal 1983 test procedure, i.e. a non-repetitive driving cycle that had been developed in the US.

After considerable discussion the standards shown in Table 1.2 were agreed (the 'Luxembourg Agreement') by most Member States. However, Denmark and Greece at the time of writing had not agreed and had entered reservations. It is of interest to note that the basis for the emission limits has been changed from vehicle weight in the Regulations up to 15-04 to engine size. Thus the most stringent requirement is for the largest engine size range (2.0 litres). Furthermore the emission levels refer to the ECE R15 driving cycle. At the time of writing the European Commission is coordinating work in the homologation laboratories of Member States and industry to evaluate proposed extensions to the ECE

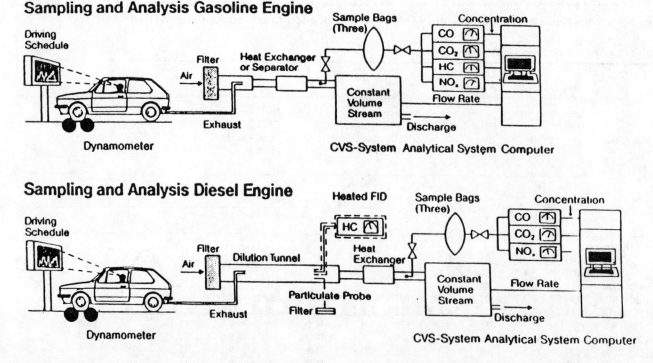

Figure 1.7 Schematic Arrangement of Constant Volume Sampling Systems (CVS) for Gasoline and Diesel Engined Vehicles

R15 drive which would include rural and highway driving simulations. It is considered by some major manufacturers that these extensions to the test cycle could influence the technological decisions for emissions control. Thus, for example, if the limits set for nitrogen oxides emissions at the higher cycle speeds are too stringent then three-way catalyst control may need to be introduced over a wider product range.

Table 1.2 The Luxembourg Agreement Proposed Limits for Light-Duty Vehicles

Cylinder capacity (litres)	Mass of carbon monoxide (g/test)	Combined mass of hydrocarbons and nitrogen oxides (g/test)	Mass of nitrogen oxides (g/test)
>2.0	25	6.5	3.5
1.4 to 2.0	30	8	–
<1.4	45	15	6

International Test Cycles for Light Duty Vehicles

The ECE R15: The urban driving cycle is shown in Figure 1.8. The average speed of this cycle is 18.7 km h^{-1}. The distance per cycle is 1.013 km but a test comprises four cycles resulting in a total distance of 4.052 km.

The US Federal 1975 Driving Schedule: This drive is a simulation of a commuting trip in Los Angeles. There are three phases: a cold start phase of 505 seconds duration with a maximum speed of 56 mph, a 'stabilised' phase and a hot start phase in which the vehicle is driven to the same speed/time pattern prescribed for the cold start phase. The speed/time profile is shown in Figure 1.9.

The average speed of this cycle is 31.7 km h^{-1}. The test distance is 17.8 km.

The Japanese Cycles: The Japanese 10 and 11 mode driving cycles are shown in Figure 1.10. The 10 mode cycle has a driven distance of 0.664 km per cycle; it is a hot start 'stabilised' test in which the cycle is repeated six times to give a total distance of 3.98 km.

The 11 mode test cycle is repeated four times to give a total distance of 4.08 km per test. In this case the test is a cold start procedure.

THE EUROPEAN SCENE FOR HEAVY DUTY DIESEL ENGINES

The European Commission (CEC) has proposed limits and test methods for the Type Approval of heavy-duty diesel engines. The proposed Directive is based on the ECE Regulation 49 which, however, has not been adopted by Member States of the EEC. Whereas the EC proposal was not finalised at the time of writing of this Manual, it has been agreed that the EC requirements for gaseous emissions should result in a reduction of 20% for CO and NO$_x$

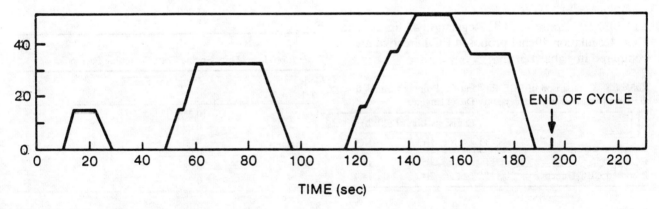

Figure 1.8 ECE R15 Urban Driving Cycle

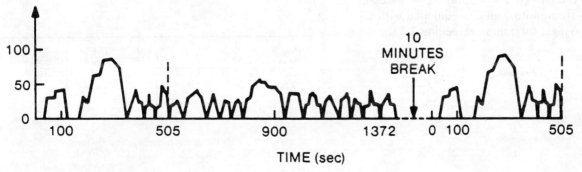

Figure 1.9 US Federal 1975 Driving Schedule

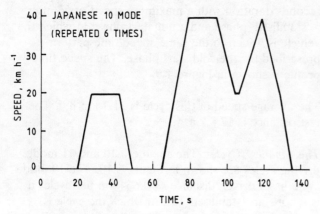

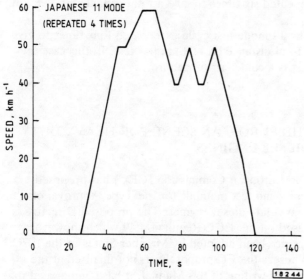

Figure 1.10 Japanese 10 and 11 Mode Driving Cycles

and 30% HC from the ECE Regulation 49 limits. The ECE Regulation 49 and proposed EC Directive are compared in Table 1.3.

Table 1.3 Comparison of ECE R49 and EC Proposed Limits for Gaseous Emissions from Heavy-Duty Diesel Engines

	Emissions Limit (g/kWh)		
	CO	HC	NO_x
ECE R49	14	3.5	18
Proposed CEC Directive	11.2	2.4	14.4

Heavy duty diesel engines are defined as engines which work on the compression-ignition (CI) principle; the Directive will apply to the gaseous pollutants from motor vehicles equipped with CI engines having a total mass exceeding 3.5 tonnes and excluding vehicles which run on rails, agricultural tractors and machines and public works vehicles.

The test is carried out on an engine test bed equipped with a suitable dynamometer and associated measuring instruments such as instruments to measure engine speed, torque, fuel consumption. Ambient test cell operation conditions are also measured (e.g. atmospheric pressure and humidity).

The test cycle has 13 conditions or modes which are shown in Table 1.4. The emissions in each mode are measured and calculated on a mass basis and weighting factors are applied for each mode. The proposed weighting factors are given in Table 1.5.

The second stage of controls on heavy duty diesel engines will probably involve particulate limits. However, the timetable for these discussions had not been proposed at the time of this Manual.

Table 1.4 Proposed CEC 13 Mode Cycle for Heavy-Duty Engines

Mode No.	Engine Speed	% Load
1	idle	–
2	intermediate	10
3	intermediate	25
4	intermediate	50
5	intermediate	75
6	intermediate	100
7	idle	–
8	rated	100
9	rated	75
10	rated	50
11	rated	25
12	rated	10
13	idle	–

Table 1.5 Heavy-Duty Engine Test: Weighting Factors

Mode No.	Weighting Factor
1	0.25/3
2	0.08
3	0.08
4	0.08
5	0.08
6	0.25
7	0.25/3
8	0.10
9	0.02
10	0.02
11	0.02
12	0.02
13	0.25/3

2. Emissions Control Technologies—Spark Ignition Engines

2.1 Fuel Systems, Ignition Systems and Control Devices—Principles

2.2 Current Engines

2.3 Lean-Burn Engines

2.4 Catalytic Systems

2.5 Application of Control Technologies

2. Emission Control Techniques – Spark Ignition Engines

2.1 Fuel Systems, Ignition Systems and Simple Control Devices – Basic Principles

The pollutants from gasoline engines are shown in graphical form in Figure 2.1. It will be seen that the concentration is greatly influenced by the air/fuel ratio of the mixture that is drawn into the engine and subsequently burned in the combustion chamber. The dotted line indicates the stoichiometric ratio of air to fuel which is the exact chemical quantity of air required to burn the fuel completely. Unfortunately due to the difficulty of perfect mixing in the very short time available, combustion is not quite 100%. On the left of the dotted line there is excess fuel and insufficient oxygen from the air and hence hydrocarbons and carbon monoxide are relatively high, for leaner (i.e. less fuel) settings both these pollutants are reduced but not eliminated. Nitric oxide which is formed in the combustion chamber from the nitrogen and oxygen in the air, due to the high temperature of combustion reaches a maximum just weak of the stoichiometric A/F ratio. At richer (i.e. more fuel) A/F ratio there is insufficient oxygen, and as settings are leaned off the temperature of combustion is reduced and less nitrogen and oxygen combine. As the nitric oxide formation increases with the temperature of combustion the control of the ignition point by the spark is critical. Retardation reduces the temperature and hence the NO_x but unfortunately also reduces the efficiency of combustion with consequent deterioration of fuel consumption. On the right of the dotted line (Figure 2.1) it can be seen that by weakening the mixture to A/F ratios beyond about 16:1 substantial reductions in NO formation are achieved.

It will be appreciated from the above discussion that the accurate control of A/F ratio and spark timing are therefore vital for the reduction of pollutants.

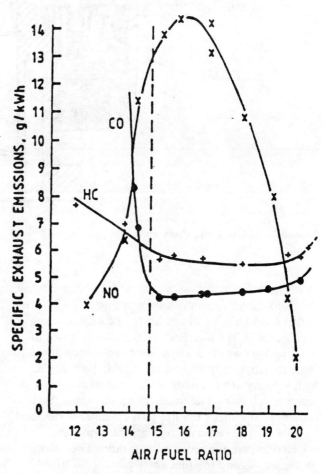

Figure 2.1 Variation of Emissions with Air/Fuel Ratio

very important part of this enquiry and it is necessary to examine the various systems in some detail.

The function of a fuel system is to provide an engine with a mixture of fuel and air in the correct proportions and, as far as possible, in equal quantities to each cylinder. Additionally the fuel charge when it reaches the time of ignition should be fully evaporated and mixed with the air. This involves great sophistication in the design of the inlet manifold and carburettor. Owing to the pulsating nature of a reciprocating engine this is a complex problem.

Carburettors

There are two main types of carburettor, the fixed choke and the variable choke. The arrangement of

FUEL SYSTEMS

All of the following remarks apply to 4 stroke cycle engines. It has been observed (Figure 2.1) that the A/F ratio is the principal parameter that characterises the pollution pattern and in particular nitric oxide. It is thus of the utmost importance that this is controlled with precision. The fuel system is thus a

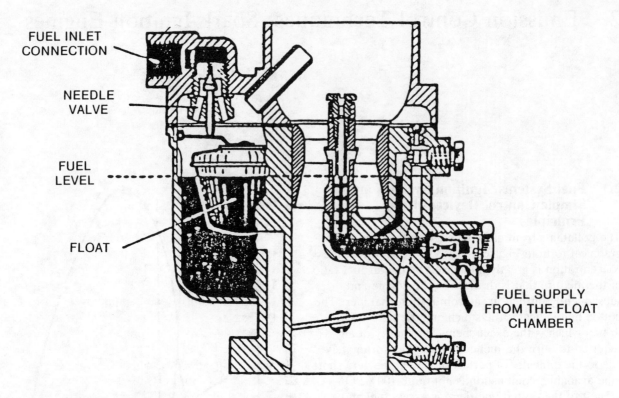

Figure 2.2 A Fixed Choke Carburettor

the fixed choke system is shown in Figure 2.2. Air is drawn through the carburettor by the engine being accelerated in the so-called 'choke tube' (shown in the diagram) which is a venturi, sized sufficiently large to avoid restriction to the engine power. The high velocity at the throat of the venturi reduces the pressure, allowing fuel to be drawn in from a float chamber. The fuel from the float chamber is metered by 'jets' that have small, very accurate orifices, selected to give the required A/F ratio. The velocity of the air passing the fuel entry point, 'the beak', atomises the petrol allowing it to be suspended as droplets in the air and it is then conducted through the inlet manifold to the cylinders.

Special provision has to be made for acceleration when the throttle is suddenly opened. A so-called acceleration pump is used for this purpose and consists of a simple plunger that is connected to the throttle linkage and supplies a jet of fuel into the air stream each time the throttle is depressed.

For cold starting 'a choke' (not to be confused with the choke-tube) consisting of a butterfly valve is used; this causes a high depression at the beak and draws in considerable extra fuel to compensate for the fact that under cold conditions the heavy fractions of the petrol do not evaporate, and the effective A/F ratio would be so high that it would not ignite. However these heavy fractions which pass through the engine, only partially burned, are a major source of hydrocarbon pollution.

As will be appreciated, a richer mixture is required for acceleration and maximum power than for cruise and for this purpose compensating jets are employed.

The variable choke carburettor is so named because it provides a choke-tube having an orifice size varying with the requirements of the engine. This gives an approximately constant air velocity and hence pressure at the fuel entry point providing a better degree of fuel atomization than is possible with a fixed choke at low engine speeds.

The mode of operation is illustrated in Figure 2.3. The size of the orifice is varied by the vertical movement of a piston (1) positioned above the fuel jet (2). A suction disc (3) integral with the piston operates in a cylinder. Drillings in the piston transmit the depression existing in the duct between the throttle disc (5) and the piston thus raising the piston and increasing the choke size. When the throttle is opened the engine accelerates and takes more air, this tends to increase depression, thus lifting the piston to restore the original desired depression. The fuel orifice is varied in size in accordance with the piston position by means of the tapered jet needle (8) fixed to the piston, which is tuned to give the required A/F ratio.

A richer mixture for acceleration is obtained by slowing down the rise of the piston by a hydraulic damper, the higher consequent velocity drawing in more fuel. To enable the fuel to be evenly distributed

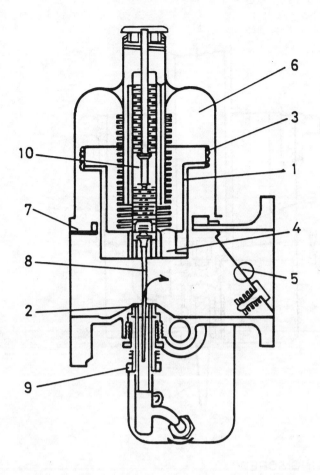

1. CLOSE FITTING PISTON
2. FUEL JET
3. SUCTION DISC
4. DEPRESSION TRANSFER DRILLINGS
5. THROTTLE DISC
6. SUCTION CHAMBER
7. ATMOSPHERIC VENT PASSAGE
8. JET NEEDLE
9. MIXTURE ADJUSTING NUT
10. HYDRAULIC DAMPER

Figure 2.3 A Section Through an S.U. Carburettor Showing the Variable Choke Components

from cylinder to cylinder, through bends and junctions, a small particle size is required so that the droplets remain airborne. This is not easy to achieve and the walls of the inlet manifold tend to become wet. To inhibit this wetness it is usual to provide heat from the hot exhaust manifold or the cooling water. The amount of heat that can usefully be supplied to the intake tracts is limited as the heating of the mixture reduces the density and consequently the mass of air drawn in and as a result the maximum torque is reduced.

A similar type is the Stromberg CD (Constant Depression) carburettor shown in Figure 2.4. The principle of operation is the same but the air valve and fuel jet are lifted by a diaphragm rather than a piston in a cylinder, as is the case in the SU design.

Having considered the basic systems of carburation which have been in use for many years and which are adequate, with precise machining, for systems to meet ECE R 15-04 regulations, the problems involved in meeting more stringent regulations will now be examined.

Throttle Body Injection (TBI)

A limitation with the carburettors described above is the range and speed with which changes can be made to meet the varying and transient conditions experienced on the road and this has led to the design and development of 'Throttle Body Injection'. The principle is that of the fixed choke carburettor but with fuel being injected at comparatively low pressure into the venturi. The fuel is metered with a solenoid operated valve which holds a precisely calibrated slit or orifice for a predetermined time. This time of opening has to meet the demands of the engine and is served by a microprocessor which receives appropriate signals from sensors. Clearly this gives a far wider range of control than the simpler systems.

The same principles have been applied to the constant depression type of carburettor but are usually considered to be unnecessarily expensive.

One such system in production works on the method outlined above, Figure 2.5. The injector is fed from a fuel pump at a pressure of 1 bar (14.5 lb in^{-2}) and is controlled by a pressure regulator mounted in the throttle body unit. The excess fuel not required by the engine spills back to the tank. The amount of fuel fed to the injector to meet engine requirements is controlled by the time the injector is kept open (pulse width); this is varied by the signal from the Electronic Control Unit (ECU). For a four cylinder engine the injector is opened twice per revolution so that each cylinder gets a pulse of fuel for its induction stroke.

Air is metered by calibrating the engine over its speed range for a series of throttle settings. This is stored in the memory of the ECU and modulated for air density changes by an ambient air temperature sensor.

Fuel is metered to provide the required A/F ratio from a map obtained in the first place experimentally on the test-bed to meet all load and speed requirements. This is then stored in the ECU. This fuel control is modulated for cold starting. The idler speed is controlled by a stepper motor which sets the throttle at the desired position to allow enough air/fuel for steady idle running even when auxiliaries such as air conditioning and power steering are operated. The total system includes Programmed Ignition within the same ECU, Figure 2.6.

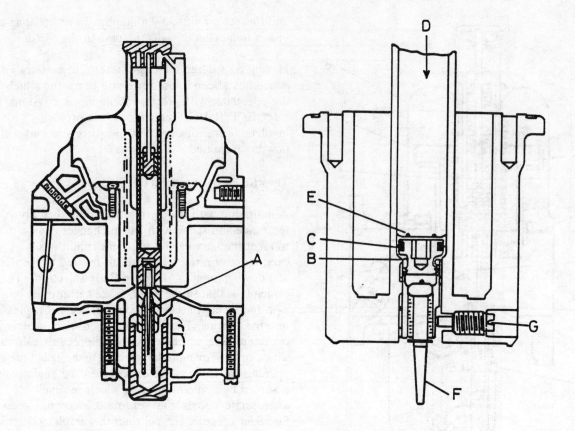

A = MAIN JET
B = NEEDLE ADJUSTING SCREW
C = "O" RING FOR ADJUSTING SCREW
D = OIL RESERVOIR
E = RETAINING WASHER
F = METERING NEEDLE
G = SPRING LOADED LOCATING SCREW

Figure 2.4 The Stromberg CD3 Carburettor

Multipoint Injection

The difficulty in metering the fuel evenly to each cylinder has been partly overcome by a further more expensive system, namely multipoint injection, in which fuel is metered and injected into the eye of the inlet valve with a low pressure injection system. The injected fuel is usually deposited simultaneously at the eye of the inlet valve and is induced into the cylinder as each inlet valve opens in turn. However, as an additional complication, the injection period may be phased to deposit the fuel at each inlet valve in the normal firing order, that is the order in which each sparking plug receives its spark. This has the advantage that fuelling conditions for each cylinder are the same.

It may at first sight seem that the multipoint injection system is ideal but it too has its problems; namely that a much smaller fraction of fuel has to be metered accurately and this, at low speeds and loads, is very difficult.

These matters were discussed with several suppliers of fuel systems, namely *Austin Rover*, *Bosch* and *Lucas* and to bring out the former points some of these modern systems will be described.

Figure 2.7 shows a block diagram of the *Lucas MPI* system. An electrically operated pump delivers fuel from the tank to a pressurised rail, where it maintains a constant pressure, the excess fuel being returned to the tank via the pressure regulator. Each fuel injector is fed from this rail. The quantity of fuel for each prevailing engine condition is metered by controlling the frequency or duration of the electrical signals to the solenoid of each injector. It will be noted that there are several sensors (left-hand side, Figure 2.8) feeding information through an analogue/digital converter to a microprocessor from which signals are conveyed to the various actuators and instruments shown on the right.

In order to control the A/F ratio it is necessary to

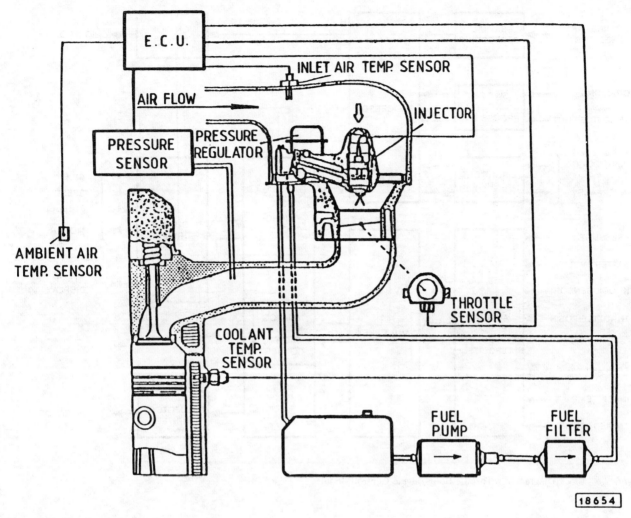

Figure 2.5 The Austin Rover Throttle-Body System

measure the air flow so that the fuel supply may be tailored to meet the requirements under all conditions. In the Lucas system this is measured by a hot wire air mass flowmeter sited between the air filter and throttle body. The sensor of this meter feeds a signal back to the electronic control unit (ECU).

The basic steady state fuelling required is read off from a three dimensional map stored in the ECU microprocessor memory with sensor feeds from the air meter (indicating load), and engine speed from a flywheel position sensor. Acceleration enrichment is initiated by sensing the throttle movement. Many other secondary controls are also sensed, which have an important significance for pollution control, in particular, coolant temperature monitoring during warm-up determines the minimum enrichment necessary. Fuel cut-off during over-run reduces hydrocarbon emissions. When 3-way catalysts are employed (see later) it is necessary to run at stoichiometric A/F ratios and the above open loop system is closed by feed-back from an exhaust gas oxygen sensor, which detects deviations from the stoichiometric A/F ratio.

Bosch have specialised for many years in fuel injection equipment, primarily to improve power for upmarket vehicles. This placed them in a good position to meet the demands for low pollution control systems. Their simplest system is the K-Jetronic and they have a more sophisticated system the L-Jetronic; both of these have derivatives known as the KE and the LH.

The K-Jetronic system is shown in Figure 2.9 and includes an electric fuel pump which generates the constant rail pressure for feeding the injection nozzles, which are located in the inlet manifold. These spray onto the back of the inlet valves of each cylinder. Air is metered by a mechanical air flow meter in the form of a pivoted vane in the air stream. This vane converts some of the kinetic energy of the stream into pressure which deflects the vane against a hydraulic force in the system; a mechanical linkage controls a spool valve which meters fuel through slits. A hydraulic counterforce acting on the control valve permits mixture enrichment during warm-up and for maximum power.

The L-Jetronic system is shown in Figure 2.10. It is an updated system with many of the features of the

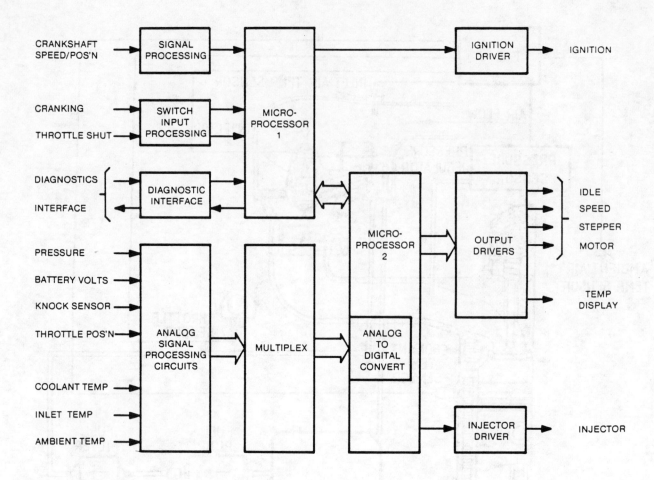

Figure 2.6 Block Diagram of a Throttle Body Engine Control Module

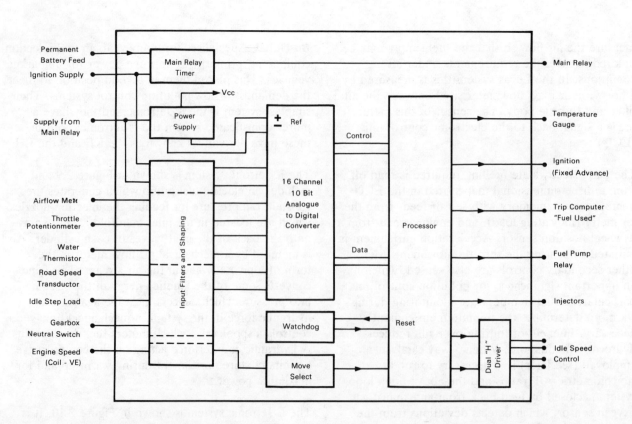

Figure 2.8 Lucas Electronic Control Module

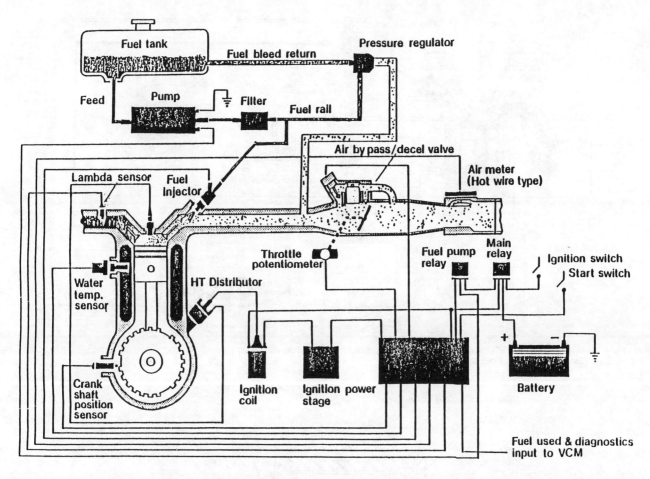

Figure 2.7 Lucas Engine Management System: 6 Cylinder

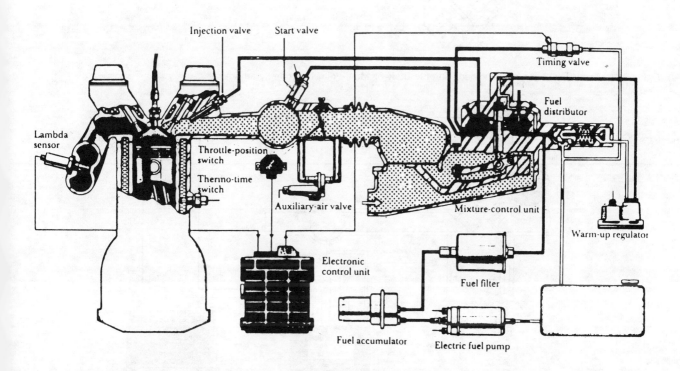

Figure 2.9 Bosch K-Jetronic Fuel Injection System

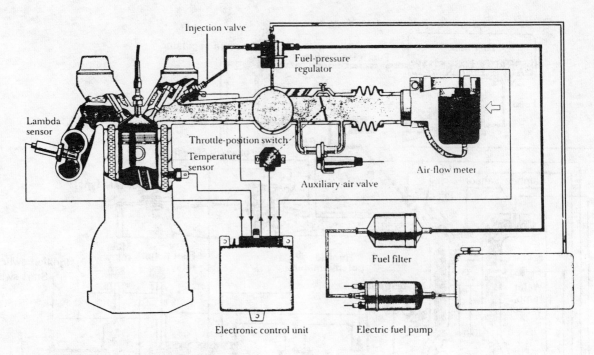

Figure 2.10 Bosch L-Jetronic Fuel Injection System

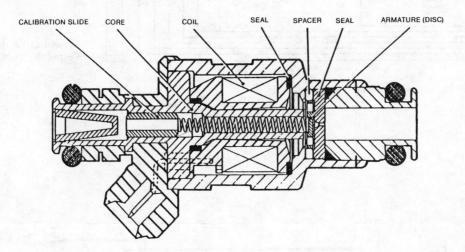

Figure 2.11 Lucus Fuel Injector Suitable for Throttle Body or Multi-Point Injection Systems

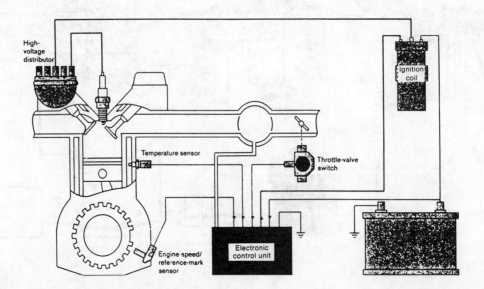

Figure 2.12 Schematic Arrangement of an Electronic Ignition System (Bosch)

K System but instead of the mechanical linkage between the vane type air-meter and the fuel metering valve, the spindle of the air-meter operates an electrical potentiometer which feeds into the ECU controlling the period of opening of the solenoid operated injectors–thus varying the fuel quantity delivered.

In the case of the LH Jetronic, a further refinement of the system has been introduced in which the air flow is measured by a hot-wire air-meter. This gives more precise measurement as it senses mass flow rather than the product of velocity times the square root of density with the vane type of meter.

An example of an injector is shown in Figure 2.11. This is the *Lucas* disc-type which may be used for single point or multipoint systems. It is operated by a fast response solenoid and is thus capable of responding to electronic signals from a fuel management system. The fuel pressure is 2–3 bar and produces, through a calibrated orifice, a fuel pencil jet or atomised cone as required by the application.

IGNITION SYSTEMS

The ignition system also plays a large part in controlling pollution and particularly nitric oxide. Because nitric oxide formation increases with the temperature of combustion, the control of the ignition point by the spark is critical. Retardation of ignition reduces the temperature of combustion and hence reduces NO formation; unfortunately retardation also reduces the efficiency of combustion with consequent increase of fuel consumption.

Figure 2.12 shows the components and general configuration of a modern electronically controlled system. However the majority of cars today (1986) have a similar system but considerably simpler and consequently less expensive. In nearly all systems the traditional sparking plug is used, which consists of two electrodes separated by an airgap of approximately 1 mm. To produce a satisfactory spark at the pressure prevailing in the engine combustion chamber, a voltage of at least 5 kV with an energy level of 100 mJ is required. This is obtained from an ignition coil which is simply an auto transformer. The primary winding is connected to the battery (12 volts), and in the conventional system a contact breaker operated by cams on the distributor drive shaft interrupts the circuit. This induces a voltage in the primary circuit which is magnified by the turns ratio to the secondary circuit which is connected to the sparking plug. Single wires are normal as the negative return goes through metallic engine parts.

A typical engine ignition characteristic is shown for full load operation in Figure 2.13 and it will be seen that the engine requires an ignition advance roughly proportional to speed. This is accomplished in the conventional system by a mechanical centrifugal advance mechanism. However for light and cruise loads an even greater advance in the ignition is required for economical operation. This is achieved, again mechanically, by a 'Suction Advance Mechanism' consisting of a suction capsule mechanically moving the contact breaker forward against the motion of the cam, thus causing the spark to occur earlier in the combustion cycle. The suction diaphragm in the capsule senses the inlet manifold pressure which indicates the engine load. This can be as much as 50° before top dead centre (TDC).

It will be seen from Figure 2.14 therefore that the optimum ignition setting depends on both speed and load and thus the three dimensional map is needed to represent fully the requirements of the engine.

Clearly the conventional system is very limited. The speed advance mechanism only allows an approximation of the ideal and the suction advance only gives a compromise for all part load conditions. Further limitations are: (1) there is backlash in the mechanical drive, (2) contact breaker cam followers wear (today these have been greatly improved), and (3) the points of the contact breaker erode. Several improved systems have been developed.

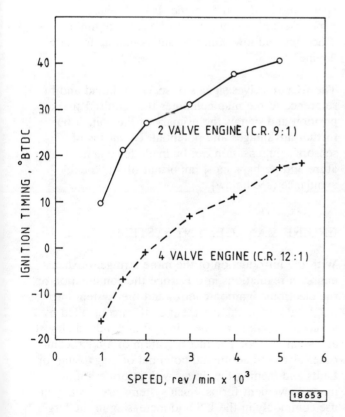

Figure 2.13 Typical Engine Ignition Timing Characteristics

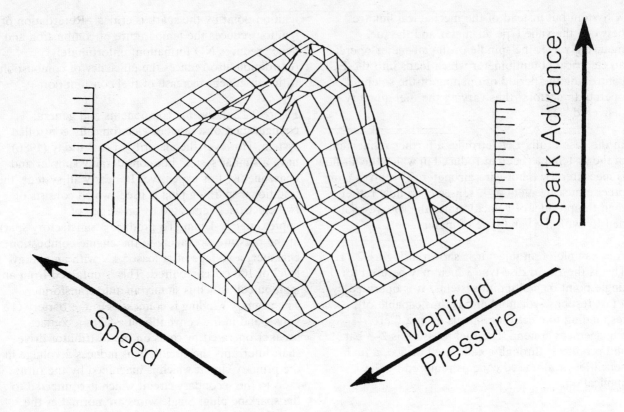

Figure 2.14 Three Dimensional Map Showing Optimal Ignition Setting Variation with Engine Speed and Load (Intake Manifold Pressure)

Transistorised Ignition Systems

This is a first step to improving the system. The contact breaker mechanism is replaced by an induction type pulse generator or a Hall-effect generator housed in the distributor as with the contact breaker in the conventional system. This signal is fed to a trigger box which triggers the primary circuit of the coil (Figure 2.15).

Various refinements may be incorporated in the trigger box. These may allow more stored energy to be derived from the coil, may control the 'dwell' affecting the duration of the spark, limit the primary current to a predetermined figure and other refinements. They still however rely on a centrifugal and vacuum advance mechanism.

Electronic Ignition Systems

A fully electronic system circumvents most of the above limitations. The distributor is now only used for distributing the spark to the various sparking plugs, the position of the piston in the cylinder is referenced from a large slotted disc which may be part of the flywheel. Again an inductive or Hall-type sensor may be used, Figure 2.16. This signal feeds an ECU giving it an accurate reference point in the engine cycle.

The 'brain' of the ECU is a stored engine map depicting the optimum ignition setting for every speed and load of the engine. It is obtained experimentally on engine test beds for optimum efficiency and low pollution and is unique for each engine.

The ECU receives signals of speed and load and by reference to the map computes the required spark position and signals the primary of the coil. It has further advantages and potential. By means of sensors compensation can be made for engine temperature and perhaps most important of all 'knock' avoidance (see below).

ENGINE MANAGEMENT SYSTEMS

With the introduction of the more stringent exhaust emission regulations into Europe the combination of the electronic ignition control and fuel system into one total engine management system, controlled by a microcomputer, will be a logical and indeed almost an essential step. The incorporation of a microcomputer gives the additional potential of diagnosing for faults and feeding the driver with information in more convenient forms. Such systems are already in use commonly in the US and increasing in the larger cars in Europe.

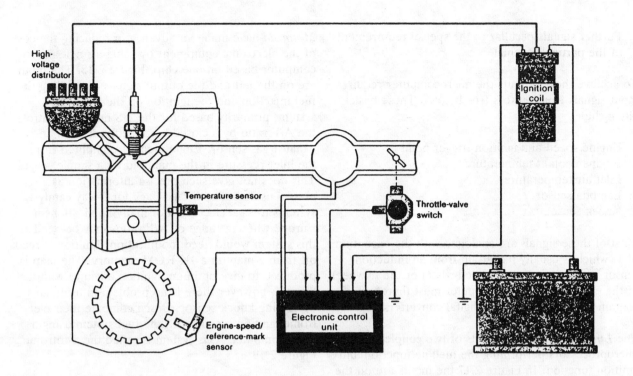

Figure 2.15 Operating Principle of Transistorised Ignition (Bosch)

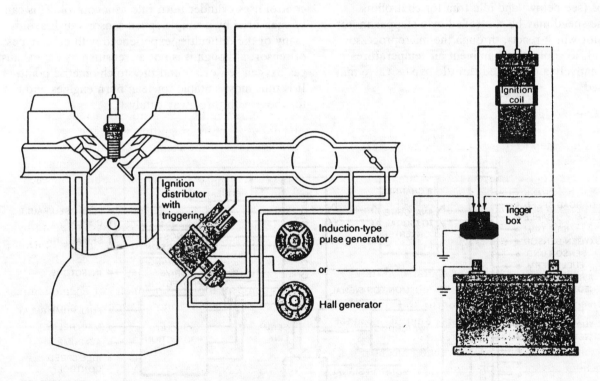

Figure 2.16 Operating Principle of Electronic Ignition Systems (Bosch)

Further developments to include controlled gear shift with automatic transmissions are being investigated with the ultimate possibility of entirely removing the direct control of the power system from the driver–the so-called 'drive-by-wire' system. This would allow the driver to choose his vehicle speed and the engine would be controlled fully automatically for minimum emissions and optimum fuelling for economy.

In order to control the engine the essential outputs required from the microcomputer are:

1. Drive signals for the fuel system
2. Drive signals for the ignition system
3. Signals for instruments, e.g. temperature gauge, engine speed, etc.
4. Idle speed control
5. Diagnostic protection systems

6. Further signals peculiar to the special requirements of the particular vehicle.

To achieve these outputs the microcomputer requires input signals from sensors (see below). These basically include:

1. Engine speed and ignition trigger point
2. Engine coolant temperature
3. Inlet air temperature
4. Lambda sensor
5. Knock sensor

Most of these signals are analogue with the exception of 1, which is usually picked up from an induction sensor in proximity to a toothed wheel or the flywheel of the engine. The microcomputer must therefore contain several analogue to digital converters.

The *Lucas* EM system consists of two coupled microprocessors performing the fuelling function and ignition function. In Figure 2.17 the inputs are on the left of the diagram and the outputs on the right.

Air flow is sensed by a hot-wire anemometer type sensor (see below) and idle trim for controlling engine speed may be controlled by a stepper motor actuator which reacts, through the microprocessor control, to coolant and ambient air temperatures. Alternatively a controlled throttle bypass valve may be used.

Motorola have made an advance in reducing the cost of the electronic equipment by using a single microcomputer based on one chip, Figure 2.18. The inputs are on the left and the output signals controlling the fuel injection and the ignition on the right. The system, primarily based for the US market, controls the A/F ratio by a closed loop feedback from an exhaust gas sensor. Engine load is computed from the inlet pressure in the manifold by a simple sensor. This does not give such precise information as air mass sensing but with feed-back for 3-way catalyst systems it is satisfactory. For many applications in Europe where a range of A/F ratios may be needed this system would need modification. Ignition is read off from a map on a 1K ROM memory. The map is produced to give optimum ignition advance without 'knock'; however there is no problem in using an overriding knock sensor. *Bosch* combine their electronic ignition system with the fuel system using a microcomputer. This system is called the 'Motronic', Figure 2.19.

More advanced feed-back type of controls are also under development. One important example of these being done by Lucas and termed adaptive engine control uses cylinder burn rate as a control. This can be measured by a simple speed sensor which avoids many of the difficulties experienced with other types of sensor, although it is not as sensitive as the exhaust gas oxygen sensor around the stoichiometric point. It is thus most suitable for lean burn engines and not for those with three-way catalysts.

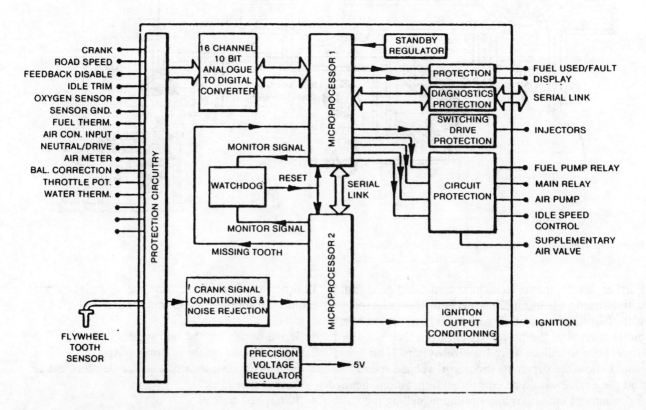

Figure 2.17 Block Diagram of Lucas Engine Management System

This system has been applied in the first place to ignition controls in such a way that it modifies the microprocessor control map to optimise the characteristics to suit the particular engine to which it is fitted.

Figure 2.20 shows the results of fitting this system to four similar 1.6 litre cars in place of a digital mapped ignition control system. The variations in the individual control maps show how engine variations were accommodated, giving an average improvement in fuel consumption of 7.6%.

While there was an increase in the average of NO_x of 4%, there was a significant reduction in the spread of the NO_x levels and, particularly, a 14.4% reduction in the highest NO_x result.

The principle is now being applied to control of fuelling and of EGR and this offers considerable prospect of improved control of NO_x in the lean burn regions. This development should reach the production stage in about two years for the adaptive ignition control and reasonably soon thereafter for the total system.

EXHAUST GAS RECIRCULATION

Exhaust Gas Recirculation (EGR) is a simple method of reducing oxides of nitrogen. Carbon dioxide is a major constituent in the exhaust gases. There is also a reasonably large proportion of water vapour derived from the hydrogen in the fuel and oxygen in the combustion air. These two gases have high specific heats and these properties enable the exhaust gas to be used to reduce the temperature of the combustion process in the cylinder. The effect of EGR is shown in Figure 2.21 for work on a single cylinder research engine. It will be noted that there are two sets of curves: one for a 9:1 compression ratio (CR) engine (conventional) and one for a 12:1 CR lean-burn engine, both at part load.

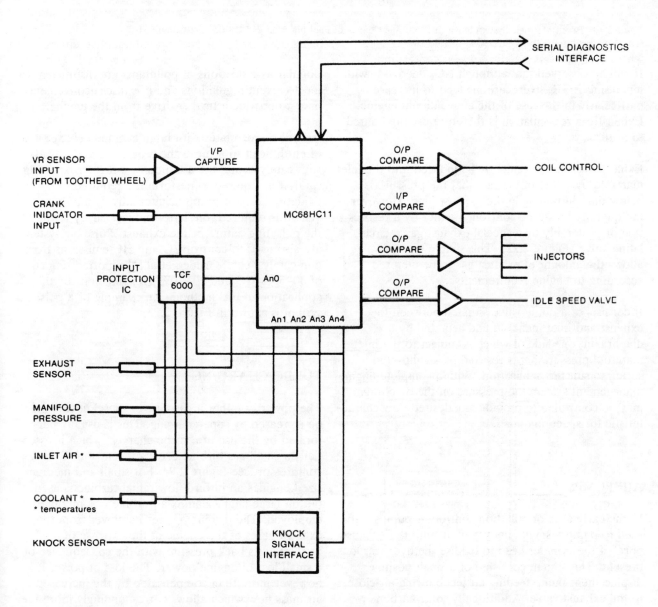

Figure 2.18 Block Diagram of a Motorola Microcomputer Based on a Single Chip

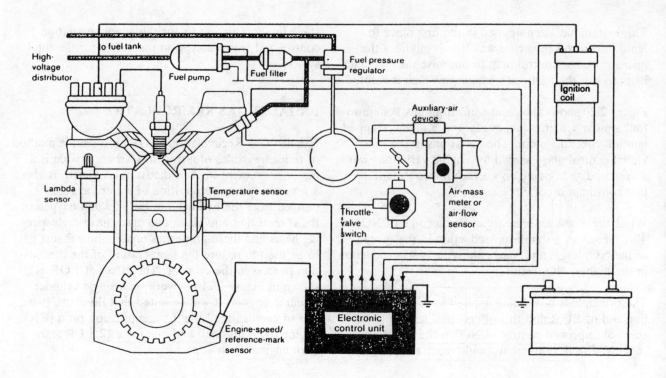

Figure 2.19 Bosch Motronic System in which Electronic Control of Fuel Injection and Ignition are Combined

It will be observed that although NO_x decreases with increasing EGR, hydrocarbons tend to increase, particularly in the case of the conventional engine. Exhaust gas recirculation is therefore usually limited to 5–10%.

Exhaust gas can conveniently be induced into the inlet manifold by virtue of the fact that the pressure is below that pertaining in the exhaust pipe. However the quantity has to be controlled either by a restriction or preferably by a special exhaust gas recirculating valve (Figure 2.22). This simple type of valve allows the amount of exhaust gas to be metered according to engine requirements.

It consists of a pintle valve situated between the exhaust and inlet manifold and actuated by a diaphragm, one side of which is subject to the inlet manifold pressure but is resisted by a calibrated spring to suit the application. With a complete engine management system the pressure on the diaphragm may be controlled to provide any desired flow characteristic for speed and load.

AIR PUMPS

In the early days of pollution control air pumps were used to pump a small quantity of air into the exhaust ports of the cylinder head to oxidise the hydrocarbons present. The system consisted of a small positive displacement pump feeding air jets blowing air close to the exhaust valve. Additionally some carbon monoxide was oxidised. Today this method is not popular as reductions of pollutants are insufficient to meet current regulations and it is also inconvenient to have to provide a further drive from the engine.

As will be seen later with large engined vehicles it is often difficult to achieve the required standards even with catalysts and the addition of a further oxidising catalyst is required to perform the required oxidation. An air pump or alternatively a pulse-air system is required, the latter induces air by virtue of the pulsating nature of the exhaust. Pulse air systems are also used to control CO and HC emissions from non-catalyst cars. Gruden et al [1] reported reductions of 30–50% CO and 20–40% HC emissions by the application of this method. An example of a pulse-air system is shown in Figure 2.23.

TURBO-CHARGING

The power of either a gasoline or diesel engine may be increased by supercharging. This is usually performed by the use of a turbo-charger, which is essentially a small gas turbine driving a centrifugal compressor, see Figure 2.24. For small and medium sized engines an inward flow radial turbine is used but on large engines an axial turbine is sometimes employed. The turbine derives its power from the exhaust gases of the engine at the expense of increasing its back pressure with the consequence of a small loss of engine power. The loss of power is however more than compensated by the increased air mass flow which allows correspondingly more fuel to be burned.

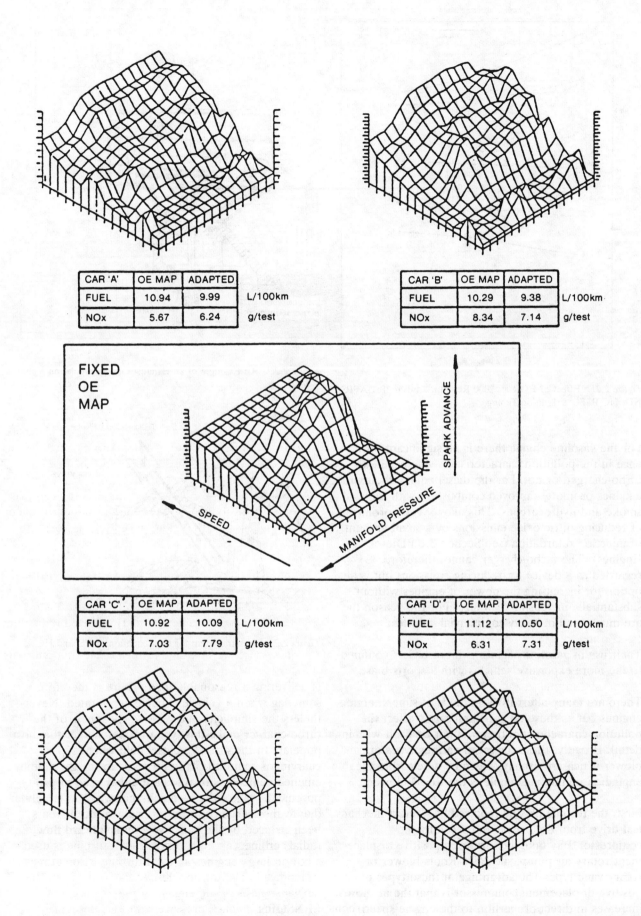

Figure 2.20 Lucas Fleet Trials–OE and Adapted Maps for Four Similar 1.6 Litre Cars

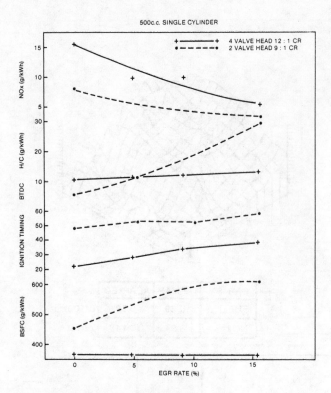

Figure 2.21 Effect of EGR at: 2000 Rev/Min 2.0 Bar 18:1 AFR NO$_x$ HC BSFC & Ignition Timing

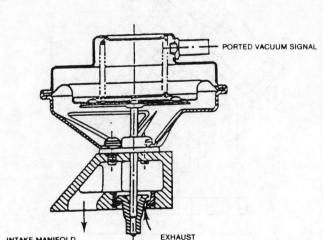

Figure 2.22 An Example of an Exhaust Gas Recirculation Valve

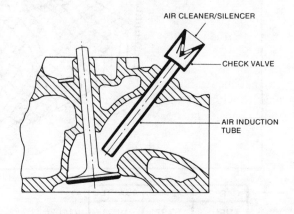

Figure 2.23 Pulsair System for the Oxidation of CO and HC

For the gasoline engine there is no significant difference in the pollution characteristics, whether it is turbo-charged or not. For the diesel engine the extra air does promote improved combustion with reduced smoke and hydrocarbons. This allows a compromise of reducing nitric oxide emissions by a small amount of injection retardation (see Section 3 on Diesel Engines). The turbo-charger cannot therefore be regarded as a device for reducing emissions but is an option for increasing the power of engines without substantially increasing their size. For this reason they are much used in heavy commercial vehicles.

Their use in gasoline engines is more or less confined to the more expensive vehicles with a sports bias.

There are many alternative methods of supercharging engines but as these do not significantly affect the pollution characteristics they will not be dealt with in detail. Broadly, these consist of various types of blower which are usually driven by a belt from the engine. A common example is the Roots blower.

First, the turbine may be replaced by a direct mechanical drive from the engine and secondly the compressor may take the form of a positive displacement rotary air pump, such as a Roots blower or rotary vane type. The advantage of these types of positive displacement compressor is that the air flow increases in direct proportion to the engine speed thereby providing a reasonably constant increase in boost pressure over the whole speed range. The turbo-charger being entirely uncoupled from the engine does not produce its boost pressure until the engine is delivering a reasonable load. There is therefore some lag when a vehicle is being accelerated. Nevertheless the simplicity and relative cheapness of the turbo-charger make this type of supercharger the most popular. In the 4-stroke diesel field where supercharging is required, it is universal. In the gasoline engined vehicles both types are either used or proposed. The lag of the turbo-charger which is largely due to the inertia of the rotor, has, in recent years, been reduced by the use of a ceramic inward flow radial turbine.

SENSORS

For the gasoline engine (as noted above under Engine Management Systems) the electrical signals required to activate and control the fuel and ignition systems

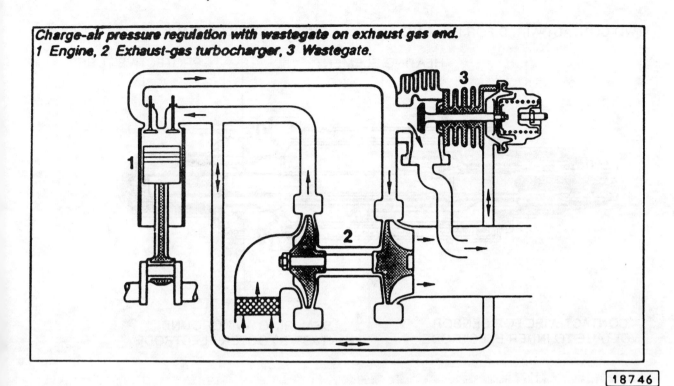

Charge-air pressure regulation with wastegate on exhaust gas end.
1 Engine, 2 Exhaust-gas turbocharger, 3 Wastegate.

Figure 2.24 An Example of a Turbocharger Installation (Bosch Automotive Handbook)

are obtained by some form of sensor. Some typical sensors are described in this section. It will be appreciated that the environment of an engine is a very hostile one. Those sensors which have to be mounted on the engine, such as the oxygen sensor and knock sensor have to withstand both heat and severe vibration. As a consequence, there has been a long and difficult period of development.

Oxygen Sensors

For three-way catalyst systems for gasoline engines (see Section 2.4) the fuel system has to supply the engine with a stoichiometric mixture of fuel (gasoline) and air (A/F ratio 14.7). The zirconia cell which is sensitive to the small amount of oxygen present in the exhaust gas has suitable characteristics for this purpose and is frequently termed a 'Lambda' sensor. 'Lambda' (λ) is defined as the ratio of the actual A/F ratio divided by the stoichiometric A/F ratio and for 3-Way catalyst this ratio has to be unity. The characteristics of the cell are shown in Figure 2.25 and a cross section of the cell itself in Figure 2.26. It will be noted from the graph that if the mixture supplied to the engine goes rich ($\lambda = 1$) a voltage of approximately 1.0 is generated. This provides a signal to the microprocessor which in turn provides a signal to the ECU which controls the duration of opening of the fuel injector thus restoring the ratio to $\lambda = 1$. The process is termed 'feed-back' control. It will be noted that there is almost a step change for a small increase or decrease in the A/F ratio about stoichiometric, thus making the system sensitive. Although this type of cell will operate without electrical heat an electric heating element is usually incorporated for stability, quick operation from a cold start, and the reduction of poisoning from contaminants in the fuel. The principal developer of this type of cell for automotive use is the *Bosch* organisation.

Titania (TiO_2) sensors are also being developed, which are claimed to have some advantage over the Lambda sensor. These are in the form of thick film semi-conductors which change their electrical resistance with oxygen potential and temperature. Temper-

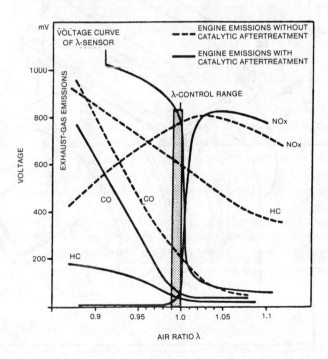

Figure 2.25 Characteristic of Zirconia λ Sensor and Exhaust Gas Concentrations

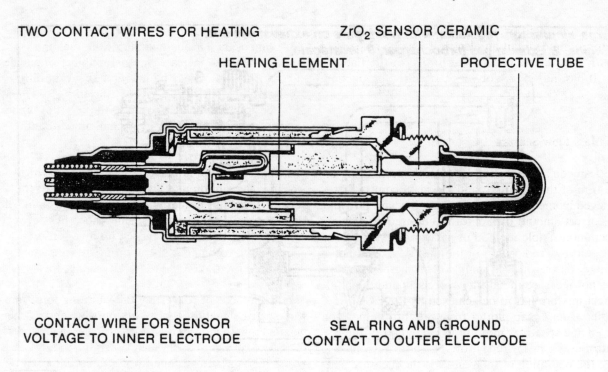

Figure 2.26 Headed Zirconia Lambda Sensor

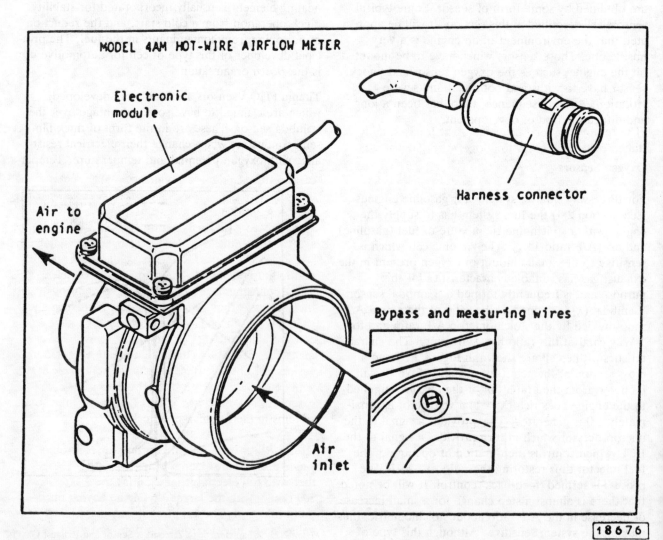

Figure 2.27 Lucas Air Mass Flow Sensor

ature is controlled by a heating element, and with an appropriate circuit an electrical voltage response to A/F can be obtained similar to that of the Lambda cell. It has, however, not yet reached the production stage.

Air Mass Flow Sensors

The Lambda sensor is only capable of performing the fine trim of the A/F ratio and the main control is achieved by measuring the air mass flow and providing the correct amount of fuel for the desired A/F ratio. A common simple method of measuring the air flow is by sensing the position of the throttle from a potentiometer on the throttle spindle. This method however does not measure the air flow precisely and the correct mixture has to be achieved by the ECU reading from a map plotter from the throttle position and engine speed. A more precise method of fuel control uses a hot-wire sensor. The principle underlying the operation of this sensor is the measurement of the rate of loss of heat from a heated wire which is a function of the velocity and density of the air passing over it. Electrical energy is supplied to maintain a constant selected temperature and the measure of this energy provides a signal for controlling the appropriate amount of fuel.

A Lucas hot-wire sensor is shown in Figure 2.27.

Knock Sensors

The removal of lead from fuel, which has already taken place in the US to allow for the use of catalysts, is following also in Europe. This results in the lowering of the Octane number of the fuel and thereby tends to increase the susceptibility of engines to 'knock'. Knock may be prevented by retarding the ignition albeit with a loss of efficiency. It is therefore desirable to be able to sense the onset of 'knock' and to retard the ignition by a small amount automatically.

Knock sensors are in reality accelerometers which pick up the vibration of an appropriate part of the engine which is excited by knocking combustion. The characteristic frequency of engine knock lies between 3–9 K Hertz and may be picked up by a sensor designed to resonate at the appropriate frequency measured by trial on the required engine.

Many methods of converting the vibrating force into an electric signal for control are possible and a schematic cross-section of a piezoelectric ceramic sensors is shown in Figure 2.28. Such a sensor produces an electric signal proportional to the forces imposed by the resonant vibration. This signal feeds into the engine management system and provides the required minimal retardation.

2.2 Current Engines

GENERAL DESCRIPTION

European vehicles now meet or are capable of meeting the ECE R15-04 Regulations. With some vehicles it may be just possible to do this with careful carburettor tuning and a contact-breaker type ignition. However most manufacturers are employing fully electronic ignition or at least breakerless systems (see Ignition Systems, Section 2.1). There are several electronically controlled single point injection systems on mass-produced cars, while most offer multipoint injection on their up-market vehicles. These refinements are fitted as selling features in themselves as they give improved performance and economy. At the same time however they have considerable potential for pollution control and make the meeting of current European Regulations easier. Ford fit electronic injection on all vehicles with single variable venturi carburettors, twin venturi carburettors, and *Bosch* K and L Jetronic injection systems in their upmarket vehicles.

Austin Rover have full microprocessor controlled single point (Throttlebody) injection systems together with full electronic ignition for the 2 litre Rover 800 range, with the larger V6 engine having multipoint injection. The 1.6 litre engine in the small vehicles has full electronic ignition, and the 1.3 litre, breakerless systems.

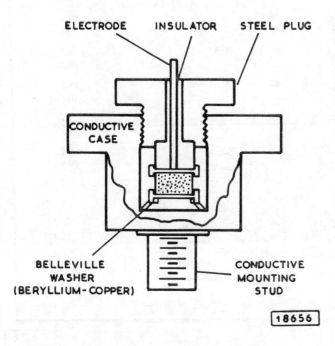

Figure 2.28 Schematic of Piezoceramic Knock Sensor

Similar approaches were found with the European manufacturers *Volkswagen* (VW), *Fiat* and *Peugeot*. *Peugeot* have breakerless ignition on all vehicles. *VW* particularly stated that inevitably the more sophisticated the fuel and ignition equipment, the more expensive it would be, and the smaller vehicles could not bear the costs. By and large with their present engines, to meet 15:04 regulations, it is *VW* policy to tune their fuel systems to give an air/fuel ratio between 16:1 and 17:1 for cruising, and to richen to 14:1 for maximum power and acceleration.

EMISSIONS AND FUEL CONSUMPTIONS

In the UK the current legislation is based on ECE R15-03. Hence up to October 1984 EC certification could be issued to that standard, and since that date only national Type Approval (TA) to that standard can be issued. However, for commercial reasons manufacturers have been applying for TA to ECE R15-04 and hence one major manufacturer has '15-04' TA for most of the model range. The implications of this legal situation are that emission levels for most new vehicles are, in practice, about the same levels as the ECE R15-04 TA or Conformity of Production (COP).

Six gasoline engined vehicles of engine size 1.0 to 2.8 litres and two diesel engined vehicles which had been homologated to ECE R15-04 were tested in accord with the requirements of this regulation. The results are presented in Table 2.1.

In the case of the gasoline cars it can be seen that one car (a 1.6 litre) exceeded the TA standard for CO and HC+NO_x, and the 2.8 l car exceeded its appropriate CO standard. These comparisons, however, should be treated with caution and are given as indicators only of the in-use performance of nearly new cars (i.e. less than 1 year old and odometer readings not greater than 10,000 miles) when tested in accord with the ECE R15-04 regulations.

The diesel engined cars were low emitters of CO and HC and relatively low emitters of NO_x. Thus the highest NO_x emission was 3.68 g test^{-1} for a 2.3 l diesel car which was 16% less than the 1.0 litre gasoline engined car. The low HC emissions and relatively low NO_x emissions resulted in combined HC+NO_x emissions that were 17 to 18% of the TA standard in the case of the 1.6 litre cars and about 25% of the appropriate standard in the case of the 2.3 litre cars.

The fuel consumptions were calculated by the carbon balance technique for comparison purposes. For

Table 2.1 ECE R15-04 Emissions Tests for In-Service Vehicles (Means of 6 Tests)

	Emissions, g test^{-1}				ECE R15-04 Standard g test^{-1}		Fuel Consumption l (100 km)$^{-1}$
	CO	HC	NO_x	HC+NO_x	CO	HC+NO_x	
Gasoline Engined Cars							
1.0 l (FWD)	58.7	11.8	4.40	16.2	58	19	8.60
(SD)	(4.6)	(0.6)	(0.4)				(0.6)
1.4 l (FWD)	49.5	10.1	7.80	17.9	58	19	10.1
(SD)	(6.9)	(0.6)	(0.8)				(0.6)
1.6 l (FWD)	31.4	9.39	7.71	17.1	67	20.5	10.9
(SD)	(6.4)	(0.64)	(0.3)				(0.4)
1.6 l (FWD)	81.7	17.6	6.86	24.46	67	20.5	11.1
(SD)	(17.0)	(2.2)	(1.0)				(0.8)
2.2 l (RWD)	52.1	9.93	11.6	21.53	76	22	13.7
(SD)	(6.6)	(1.5)	(0.6)				(0.4)
2.8 l (FWD)	85.9	10.3	8.97	19.27	76	22	18.1
(SD)	(9.3)	(1.0)	(1.1)				(0.9)
Diesel Engined Cars							Particulates, g test^{-1}
1.6 l	3.66	0.71	2.83	3.54	67	20.5	– 6.53
(SD)	(0.52)	(0.12)	(0.30)				(0.17)
1.5 l (2 tests)	4.28	0.59	3.07	3.66	67	20.5	0.35 6.68
2.3 l	5.72	2.09	3.44	5.53	76	22	– 9.68
(SD)	(0.7)	(0.2)	(0.2)				(0.1)
2.3 l (2 tests)	6.50	1.74	3.68	5.42	76	22	1.23 9.72

FWD : Front Wheel Drive
RWD : Rear Wheel Drive
SD : Standard Deviation

example, the 1.6 litre gasoline engined cars had an average fuel consumption of about 11 litres per 100 km whereas the 1.6 litre diesel engined cars had an average fuel consumption of about 6.6 litres per 100 km which was a reduction of 40% as compared with the gasoline vehicles.

In the real situation motor vehicles are operated over a wide range of conditions; hence methods have been developed to determine on-the-road emissions. In the UK, Warren Spring Laboratory developed a miniature CVS system for use on board the test vehicle, and in the FRG driving patterns have been studied and driving cycles have been constructed which include speeds of up to about 120 km h^{-1}.

The variations of emissions with vehicle average speed have been measured for WSL for a sample of twenty cars [2] during urban, suburban, rural and motorway drives. All the cars had been tuned in accord with the manufacturers' recommendations. It was found that the emissions of CO and HC were strongly dependent upon vehicle average speed. The highest CO and HC emissions occurred during urban drives in the average speed range 17 to 22 km h^{-1}. As the average speed increased these emissions reduced rapidly but at approximately 80 km h^{-1} the CO emissions started to rise again. The tendency for emissions to rise at speeds above 80 km h^{-1} was much less marked in the case of total hydrocarbons. The variation of NO$_x$ emissions with average speed was not consistent; however, emissions at 113 km h^{-1} could be similar to or up to 2.4 times the urban value.

There was a considerable spread in the results for CO, HC and NO$_x$ emissions. For clarity this spread is shown in Figure 2.29 for the three pollutants as emissions envelopes.

Figures 2.30, 2.31 and 2.32 contain plots for six nearly new cars (homologated to ECE R15-04) which were tuned in accord with the manufacturers' data. Hence a direct comparison may be made between these cars and the mix of vehicles from ECE R15-00 to ECE R15-03 used in the 20 car study.

In the case of CO Figure 2.30 shows that five cars of engine capacity 1.6 to 2.8 litres had similar and relatively low emissions under urban and suburban driving conditions, that is, at average speeds up to about 40 km h^{-1}. However, the 1.0 litre car had relatively high CO emissions at average speeds up to about 60 km h^{-1}. All six cars had emissions throughout the range of average speeds up to about 115 km h^{-1} which were within the 20 car envelope, or a little below under urban conditions (about 20 km h^{-1}). Hydrocarbons emissions (Figure 2.31) for the six cars were spread fairly evenly through the envelope with some exceeding the upper boundary. The emissions of nitrogen oxides shown in Figure 2.32 were also evenly spread through the emissions envelope with an indication of a greater increase with higher speed than observed in the 20 car study. Figure 2.30 shows that six low mileage cars built to meet ECE R15-04 regulations had NO$_x$ emissions which covered the range of emissions measured in the 20 car survey. However, under the ECE R15-04 regulations these cars were within the Conformity of Production (COP) standards. Hence it can be assumed that lower in-service emissions of NO$_x$ will be achieved only when vehicles are calibrated to a tighter international standard such as the Luxembourg Agreement.

Emissions at High Speeds

A five car study of emissions at high speeds was carried out by WSL on behalf of the Transport and Road Research Laboratory of the Department of Transport. Four cars were equipped with carburettor fuelled engines and one was equipped with open loop electronic fuel injection (EFI). All the cars were low mileage and tuned to the manufacturers' specifications. Motorway drives at an average speed of 90 km h^{-1} ($\pm$10 km h^{-1}) and an average of 105 km h^{-1} ($\pm$10 km h^{-1}) were followed by transient drives on a test track up to wide open throttle conditions. The maximum target speed was 150 km h^{-1} ($\pm$10 km h^{-1}). The data are plotted in Figures 2.33, 2.34 and 2.35.

The four carburettor fuelled cars will be considered first and then the EFI car data compared with them.

Figure 2.33 shows that above 100 km h^{-1} the CO emissions of the four carburetted cars increased rapidly with average speed. At the maximum average speed (150 km h^{-1}) the CO emissions ranged from 11.6 (Car D) to 35.3 (Car C) g km^{-1}. Car C was equipped with a 1.6 litre engine, the smallest of the group. The lowest emission occurred with the 1.8 litre engined car. It should be noted that at 150 km h^{-1} the CO emissions from the four cars exceeded the hot-start ECE R15 urban simulation drive.

In the case of total hydrocarbons (Figure 2.34) the emissions increased as the average speed increased from 90 km h^{-1} to 120 km h^{-1}. At 150 km h^{-1} the emissions of HC from two cars (C and D) continued to increase but in the cases of Cars A and B the emissions dropped slightly or only increased a small amount. All four cars emitted significantly less hydrocarbons at high speed than in the hot-start ECE R15 drive. For example Car A emitted 2.3 g km^{-1} in the hot-start ECE R15 test and 0.79 g km^{-1} at an average speed of 150 km h^{-1}.

The emissions of nitrogen oxides (Figure 2.35) increased as the average speed increased from about 90 km h^{-1} to 120 km h^{-1}. At about 150 km h^{-1} the emissions of NO$_x$ from Cars A, B and D were approximately the same as the emissions at about 120

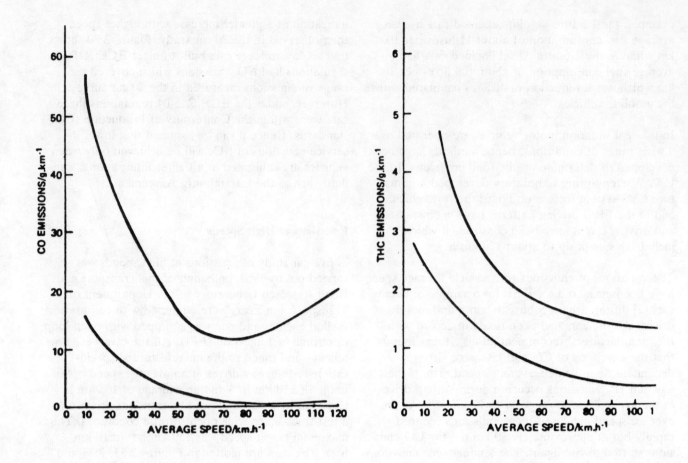

Figure 2.29 Variation of Emissions with Average Speed: Emissions Envelopes of the WSL 20 Car Survey
Figure 2.30 Variation of CO Emissions with Average Speed of In-Service Vehicles Type Approved to ECE R15-04 Compared with the WSL '20 Cars Survey' Emissions Envelope

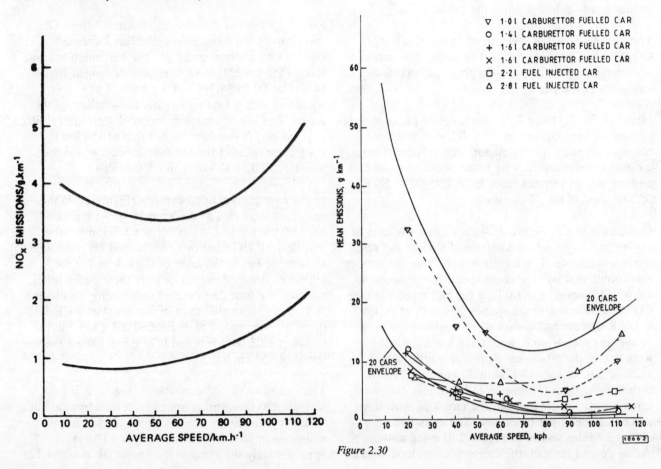

Figure 2.30

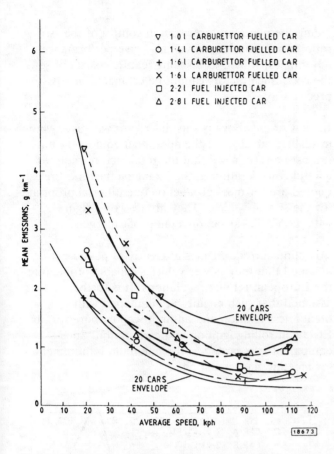

Figure 2.31 Variation of HC Emissions with Average Speed of In-Service Vehicles Type Approved to ECE R15-04 Compared with the WSL '20 Car Survey' Emissions Envelope

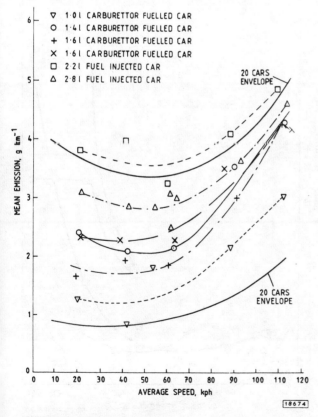

Figure 2.32 Variation of NO$_x$ Emissions with Average Speed of In-Service Vehicles Type Approved to ECE R15-04 Compared with the WSL '20 Car Survey' Emissions Envelope

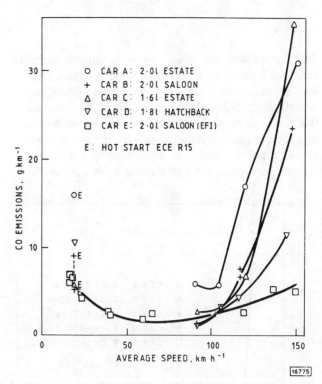

Figure 2.33 Variation of Carbon Monoxide Emissions with Average Speed

km h^{-1}. Thus the mean emissions from Car A at average speeds of 120 and 150 km h^{-1} were 3.2 and 3.3 g km^{-1} respectively; the mean emissions for Car B were 3.4 and 3.5 g km^{-1} respectively and the mean emissions for Car D were 4.7 and 4.9 g km^{-1} respectively. Car C emissions of NO$_x$ dropped from 6.6 g km^{-1} at 120 km h^{-1} to 5.2 g km^{-1} at 150 km h^{-1}. These observations are consistent with fuel enrichment for maximum power which produces high carbon monoxide emissions and reduced nitrogen oxide emissions.

The emissions of the 2.0 litre EFI car over the average speed range tested are also plotted in Figures 2.33, 2.34 and 2.35 where they may be compared with the four carburettor equipped vehicles. Figure 2.33 shows the variation of CO emissions with average speed. It can be seen that at high speeds in excess of about 120 km h^{-1} the EFI car had the lowest emissions. In the case of hydrocarbons the EFI car was also the lowest emitter at high speeds. The plot (Figure 2.34) clearly shows that relatively high emissions of hydrocarbons occur at low urban speeds in congested traffic conditions. For NO$_x$ emissions, Figure 2.35 shows that at high speeds the EFI vehicle was initially the lowest emitter but at an average speed of about 150 km h^{-1} the NO$_x$ emissions exceeded those measured with two carburettor fuelled 2.0 litre vehicles.

COSTS

Discussions with the major motor manufacturers have revealed how difficult it is to provide precise costs for the additional equipment to meet the pollution control regulations. As has been noted (Section 2.1) the fuel system is one of the most important controlling factors necessitating moving from carburettors to single point injection or some more involved system. However, the latter systems, as well as providing the necessary precision in control of the A/F ratio to minimise emissions, also give performance advantages, which, as a selling feature, partly offset the cost. Such factors cannot be estimated in any precise way.

In view of the above points this report will give, where available, the costs of the additional equipment but it must be recognised that the total picture requires careful consideration of the intangible factors. Prices quoted are for mass production quantities appropriate for the size of vehicle. They are ex-works figures, without VAT, car tax or retail profit margins.

To define current engines as clearly as possible it is assumed that they power vehicles capable of meeting the European R15-04 regulations. It is further assumed that such engines are equipped with breakerless ignition and an improved carburettor, an ECU controlling ignition and carburation. Such equipment is regarded as the minimum requirement for meeting the regulations.

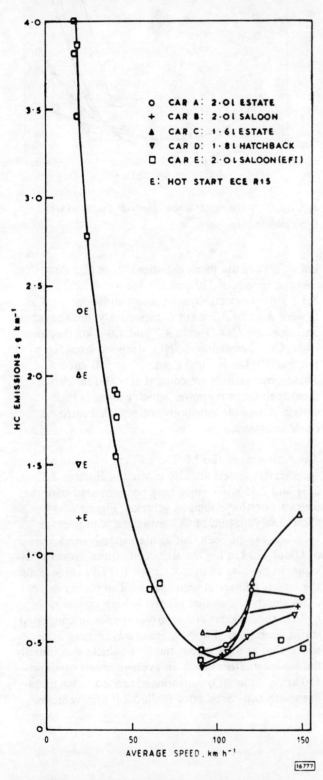

Figure 2.34 Variation of Total Hydrocarbons Emissions with Average Speed

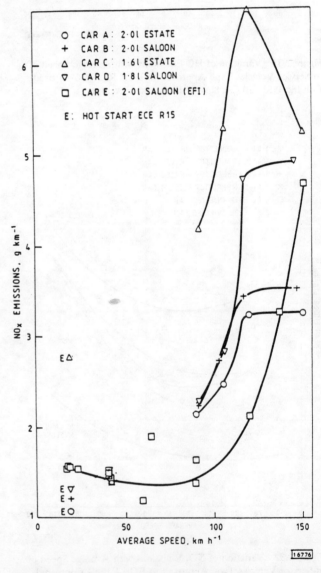

Figure 2.35 Variation of Nitrogen Oxides Emissions with Average Speed

Further improvement in driveability, power and economy may be achieved with single point (throttle-body) injection and programmed ignition, and such systems are currently fitted in higher priced vehicles. Prices for such equipment are quoted in later sections.

The base line for the additional cost of this equipment is that of a standard carburettor and contact breaker ignition which was the normal equipment for several decades before the introduction of ECE R15-02.

Additional cost for Small and
  Medium sized vehicles   c. £90) ex-works
  Large Vehicles   c. £120)

SUMMARY

The technologies applied to current vehicles produced for the European market result in emissions that meet the ECE R15-04 requirements. Careful tuning of the carburettor and basic contact breaker type ignition plus, possibly, timing adjustments will achieve these '04' limits. However, the advent of electronic ignition systems, multi-point and single point fuel injection and improved carburettors to mass produced cars have resulted in Type Approval regulatory targets being readily met. Thus, most current production cars in the UK are homologated to ECE R15-04 although UK national requirements do not go beyond ECE R15-03.

In a study of vehicles in service on the road it was shown that emissions were strongly influenced by driving pattern and average speed. In the case of CO the highest emissions occurred at the lowest average speeds, corresponding to congested traffic conditions. As the average speed increased the emissions reduced rapidly. For many cars the CO emissions started to rise again above about 80 km h^{-1}. Similar trends occurred for HC emissions but the tendency for a rise in emissions at high speeds was less marked than for carbon monoxide.

The variation of NO$_x$ emissions with average speed was not consistent. However, in most cases there was a slight initial reduction in emission as the average speed increased above the typical urban value of 20 km h^{-1} up to about 60 km h^{-1}. Above this average speed substantial increases in NO$_x$ occurred, which at 113 km h^{-1} ranged from approximately the urban to up to 2.4 times the urban value.

It is clear that the vehicle manufacturers have already made considerable progress on emission control to enable their vehicles to meet the ECE 15-04 regulations. This has had the most useful effect of fixing a baseline from which future improvements, and the cost of making those improvements, may be judged.

At a minimum, in the small popular car much improved fuel and ignition control has been introduced without a great cost to the customer and with no loss to fuel economy.

In the medium size range, generally speaking, fuel systems have received a major upgrading and ignition systems are electronically controlled. Though this is a transition stage it is expected that before more severe regulations come into force all vehicles in this range will have TBI (Throttle Body Injection) electronically controlled, with a similar control for ignition. This is being achieved with a microprocessor and appropriate sensors.

Many of the larger cars and indeed some of the medium size engined vehicles have, or will have, multi-point injection systems as a part of an integrated engine management-control package.

The expense of these systems has not been inconsiderable as indicated above and it is estimated that there has been a reduction of some 50% in carbon monoxide (CO) from the original ECE 15-00 regulated vehicles. Hydrocarbons have been reduced by a little over 25% with nitric oxide somewhat less. As the 15-04 regulations add HC + NO$_x$ together it is not possible to be precise.

It should also be borne in mind that the original 15-00 ECE Regulations concentrated efforts with the then known techniques and eliminated the bad polluters.

2.3 Lean-Burn Engines

GENERAL DESCRIPTION

As has been noted (Figure 2.1) such engines endeavour to burn weak mixtures which are homogeneous in character and as constant as can be achieved. If a large reduction in NO$_x$ is required, the A/F ratio needs to be about 20:1. Unfortunately, as the mixture is weakened the rate of burning (heat release) becomes slower and generally speaking there is a loss of efficiency. However, by raising the compression ratio and increasing turbulence in the combustion chamber, the speed of burning can be increased again, but unfortunately high compression ratios cause a greater tendency towards 'knock'. Knock is familiar to most people who can remember the low-octane fuel days during and following the Second World War when most engines 'knocked' at low speeds and high loads, namely when accelerating at full throttle, or climbing hills. This type of knock tends to die out with high speed. However, high speed knock also occurs in some, but not all, engines. It is possible to prevent this knock by ensuring that the ignition setting is prevented from advancing into

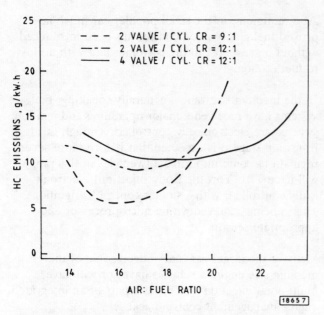

Figure 2.36 500 cm³, Single Cylinder Petrol Engine 2000 rev/min, 2 Bar B.M.E.P. HC Emissions.

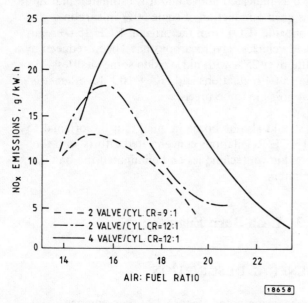

Figure 2.37 500 cm³, Single Cylinder Petrol Engine 2000 rev/min, 2 Bar B.M.E.P. No$_x$ Emissions

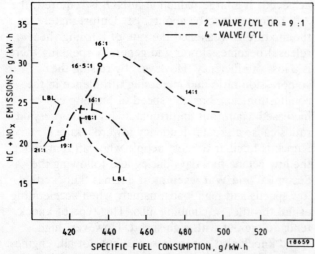

Figure 2.38 400 cm³, Single Cylinder Petrol Engine 2000 rev/min 2 Bar B.M.E.P. HC+NO$_x$ Emissions

the sensitive areas, but clearly this requires much more precise ignition equipment than is commonly used. However, such equipment is indeed available and is being progressively introduced.

Because of the great reduction and ultimately elimination of lead from fuel, the Octane Number (Research Method) of premium fuel is likely to drop from the present 98 Octane to not more than 95 and this means a compromise in the compression ratio (CR) selected. A modest CR of 9:1 could be used with a 95 Octane fuel with only a small retardation at low speeds on full load. At 12:1, about the optimum ratio for fuel economy, considerable retardation is necessary at low speeds, and some retardation at high speeds is necessary for near full-load operation. However this may be still the preferred method for small and medium sized engines to meet the new proposed emission standards (see Section 1.3). The reason for preferring this method is that at part load, under cruising conditions, knocking is not a problem and some loss of power and economy near full load is tolerated. The situation is shown in Figures 2.36 and 2.37 where pollutant concentrations are plotted against A/F ratio. The results shown are for an experimental engine running under cruise conditions with two types of cylinder head, one with a 4 valve hemispherical combustion chamber (CC) and the other with a 2 valve biscuit shaped CC. It will be noted, however, that with the 4 valve cylinder head the better filling gives an even worse peak NO$_x$ situation but combustion can be continued to a weaker A/F where low NO$_x$ is obtained. Due to the high induced turbulence with the 4 valve cylinder head fast burning is achieved with improved fuel consumption. The hydrocarbon emissions are numerically higher with the 4 valve but can be maintained to a considerably weaker A/F ratio. Figure 2.38 shows the balance of economy against pollution concentration. HC are added to NO$_x$ as in the current EEC regulations. A/F ratios are shown plotted against the concentration of pollutants. It will be seen that the lowest pollutants are obtained for the 2 valve 9:1 CR engine but this is at a considerable fuel penalty. A much improved fuel economy, with little sacrifice in pollution emission, is obtained with the 4 valve engine 12:1 CR providing it can operate satisfactorily at about 20–22:1 A/F ratio.

The discussion above is based on engine testbed research and development on a single cylinder engine and shows the potential of the lean-burn approach. When the manufacturer comes to the situation of a multi-cylinder engine in a vehicle on the road many other problems become apparent. On the fuelling side, cylinder to cylinder variation in A/F ratio can mean that some cylinders will receive higher A/F ratios than the average, and these cylinders are going to be more prone to misfire with consequent higher HC emissions. Additionally engine designers are aware of the phenomenon of cycle to cycle variations where some engine cycles show a poorer combustion characteristic from others, due largely to a variety of small differences in the initiation of combustion by the spark, plus the varying effect of exhaust products from the previous cycle. However, perhaps the greatest problems are posed by the transient conditions that prevail on the road where the engine is required to change from a full load situation for acceleration to a steady cruise condition and then perhaps overrun where the vehicle is driving the engine. These changes often take place in a very short timescale putting very great demands on the fuel and ignition systems. Against the background of these considerations the progress of the various manufacturers interviewed will now be assessed. As manufacturers are bound to address the regulations the discussion will relate to the proposed 'Luxembourg' figures which specify three engine size ranges.

EMISSIONS AND FUEL CONSUMPTIONS

The responses of 5 manufacturers of small mass produced engines (i.e. below 1.4 litres) in Europe are unanimous in that they expect to meet the regulations proposed (15:05) (the Luxembourg Agreement limits) with lean burn engines without the use of catalysts.

It is too early to say definitely what fuel systems and ignition systems will be required and whether or not EGR will be used, but from a background of considerable research and development probable configurations are proposed.

Fiat consider that some engine modifications will be needed to obtain stable running in the lean condition and that EGR will be desirable to bring the NO_x within the limits. Ignition timing will require an ECU (see above), EGR, and port liners. The latter are ceramic, or metal shrouds providing an air gap insulation, inserted in the exhaust ports to maintain the temperature of the exhaust, allowing more oxidation of hydrocarbon and carbon monoxide to take place. An ECU is required to allow 'local' retardation of ignition to avoid NO_x high spots. This is in addition to its precision advantages.

Small Engines

Austin Rover expect to meet the Luxembourg agreement level regulations with a lean-burn approach [3] and without catalysts in this size range. They also expect to include programmed ignition and may use a precision carburettor or perhaps throttle body injection.

Gaydon Technology (previously BL Technology), the research establishment of the *Rover* Group have been engaged in research and development of the lean-burn engine for some six years. Some results of this research are shown in Figure 2.39 and illustrate clearly the different response of various combustion chambers to lean-burn combustion.

Chamber	Type
A	Open 4 Valve
B	Compact
C	High Swirl
D	Disc

The advantage of the 4 valve combustion chamber is apparent – for similar NO_x + HC emission considerably better fuel consumption is obtained. If the A/F ratio is controlled at about 21:1, there is a margin in hand before misfire occurs due to mixture weakness.

A three cylinder version of the 4 valve engine was incorporated in an experimental Energy Conser-

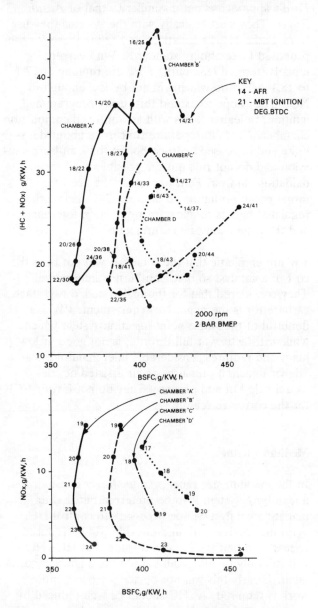

Figure 2.39 Comparison of Emissions and Specific Fuel Consumptions for Four Lean-Burn Combustion Chambers (Gaydon Technology)

vation Vehicle (ECV) and produced emission figures meeting the Luxembourg Standards, Table 2.2.

Table 2.2 Emissions from ECV During ECE R15 Cold Start Tests

	Emissions g/test			
	CO	HC	NO_x	HC + NO_x
Luxembourg Limits for less than 1.4 litre Stage I	45.0	9.0	6.0	15.0
ECV (800 kg)	8.7	10.0	3.7	13.7
ECV (910 kg)	7.9	8.6	5.2	13.8

As the ECV is lighter than a mass produced vehicle, it was retested on a chassis dynamometer with 910 kg inertial weight; it can be seen that the pollutants still met the proposed Luxembourg standards but by a smaller margin.

Ford's approach is not dissimilar to that of *Austin Rover*. They start basically with the systems they use to meet the 15.04 regulations and refine these for the proposed Luxembourg standards. With current models (termed lean-burn) *Ford* are running at 17:1 to 18:1 A/F ratio which will not be low enough in NO_x. They hope to extend this, using programmed ignition, to leaner ratios with an improved combustion chamber. *Ford (Europe)* are concerned about durability and increased hydrocarbon in the lean-burn situation and do not rule out the possible need for an oxidation catalyst. *Ford* impressed the need for target pollution figures to be some 20% below the regulated figures to allow for production tolerances and the dispersion of measurements.

VW are emphatic that the small car will not bear the cost of a catalyst so that lean-burn is imperative. They considered that for the fuel system a two stage carburettor is the minimum requirement. *VW* are doubtful of the single point injection system which, while satisfactory at full throttle, is not good at low loads and requires considerable inlet manifold heat. Higher injection pressures or air assisted nozzles would help but add expense. They do not favour ECR for the corrosive reason.

Medium Engines

In the medium size range, *Austin Rover* consider that a lean-burn system is to be preferred but that an oxidation catalyst will be necessary to control HC. With the catalyst carbon monoxide will be reduced to less than half the regulated value, but HC+NO_x will still only be marginally inside the 8 g/test requirement. Clearly this will not be adequate and more work is required. As HC should be well reduced by the catalyst, further NO reduction must be made. This may be achieved by appropriate leaning off of the A/F ratio (λ). EGR would be a further method, as indeed would be ignition retardation, but the latter would be at the expense of fuel economy. Almost certainly programmed ignition and Throttle Body Injection would be used.

Ricardo have made a contribution to the evaluation of the emission potential of engines in this range. Single cylinder research was performed on four types of combustion chamber, and one of these, the High Ratio Compact Chamber (HRCC), was chosen for full pollution investigation on a vehicle. Figure 2.40 shows pollution results for the four combustion chambers. These curves were obtained by starting with a rich mixture and weakening off until misfire occurs with attendant high HC emissions. It will be noted that a minimum HC plus NO_x is obtained with the Nebula combustion chamber, while the Bathtub, though good on NO_x, is poor on HC. The best A/F ratio for the Nebula chamber was 19.5.

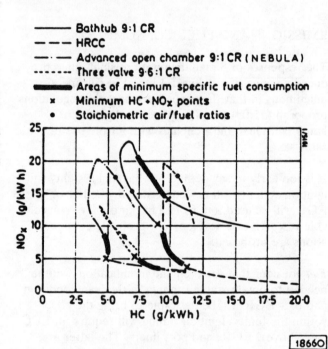

Figure 2.40 NO_x-HC Trade-Off Mixture Response 2400 rev/min 2.5 Bar B.M.E.P. (Ricardo)

The HRCC chamber (11:1 compression ratio) was incorporated in a production *VW* 1.5 l engine and installed in a car with an oxidation catalyst for pollution and economy evaluation over a range of A/F ratios. The results are shown in Figure 2.41. The Euromix fuel consumption is the arithmetic mean of the urban test and the 90 and 120 km/h test.

It will be noted, somewhat surprisingly, that the fuel consumption remains fairly constant but that the HC + NO_x as well as the CO are well reduced as the mixture is leaned off.

These tests demonstrate well the potential of the lean-burn system.

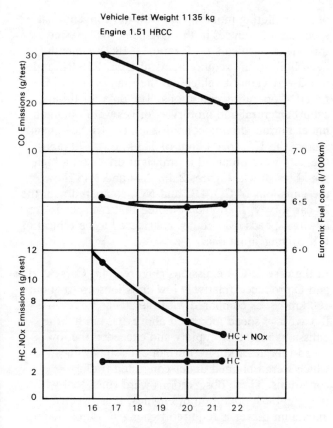

Figure 2.41 Effect of Mixture Strength on the Emissions and Fuel Economy of a 1.5 Litre Engined Car Fitted with a HRCC (CR = 11:1) Ricardo Head

Large Engines

In this size range the 'Luxembourg Agreement', regulations can be as severe as those in the US. For this reason although a lean burn approach is possible, the saving in cost is not great because of the sophistication necessary including oxidation catalysts and a complex fuel control system. The saving in fuel consumption is not considered to be justifiable at the expense of driveability. The motor industry of Europe is unanimous in concluding that they will use their well-proved systems already developed for the US market.

VEHICLE RESULTS

The emissions and fuel economies of some cars fitted with lean-burn engines when tested in accord with the prescriptions of ECE R15-04 are given in Table 2.3.

It can be seen that the evidence from *Gaydon Technology Ltd (GTL)* showed that in the less than 1.4 litre engine size class, the Stage 1 limits of the Luxembourg agreement could just be achieved by current lean burn technology without exhaust gas treatment. However it should be noted that the estimated average of 14.5 g test^{-1} of HC+NO$_x$ would imply that some cars would fail. In the case of NO$_x$ alone the

Table 2.3 Experimental and Estimated Emissions from Lean-Burn Engined Cars: ECE R15-04 Procedure

Vehicle	Exhaust Treatment	Emissions, g test^{-1}			Fuel Consumption l (100 km)$^{-1}$
		CO	HC+NO$_x$	NO$_x$	
2 l Porsche TOP	OX. Cat. + EGR	8–12	5.5–7.2	3.4–3.7	–
GTL Average‡ 1.4 l	None	30	14.5	4.5	–
GTL Average‡ 1.4–2 l	None	35	16.25	5.25	–
GTL Average‡ 2.0 l	None	35	19.0	6	–
GTL Average‡ 1.4 l	Ox. Cat.	15	7.0	4.5	–
GTL Average‡ 1.4–2 l	Ox. Cat.	15	7.75	5.25	–
GTL Average‡ 2.0 l	Ox. Cat.	15	9.0	6	–
2 l Sierra (modified)	None	12	7	2	–
	Ox. Cat.	3.5	3.5	2	–
1.6 l Toyota Carina	Ox. Cat.	6.9	5.6	4.1	–
1.6 l Toyota Carina	JMC Ox. Cat.	1.8	4.5	3.0	–
TNO: Porsche TOP (E)*	Ox. Cat. + EGR	8.6	5.4	3.8	14.0
TNO: Porsche TOP (V)†	Ox. Cat.	15.3	16.7	1.7	12.55
TNO: Passat (E)*	Ox. Cat.	16.0	3.8	1.6	10.95
TNO: Passat (V)†	None	16.4	16.6	2.4	9.85
Luxembourg Agreement Standards					
1.4		45	15	6.0	
1.4–2 litre engined cars		30	8		
2		25	6.5	3.5	

*E: minimum emissions tune
†V: minimum fuel consumption tune
‡GTL: Gaydon Technology estimations of average emissions for a broad band of engine size

estimated level of 4.5 g test^{-1} is 25% below the proposed limit of 6 g test^{-1}. However, treatment of the exhaust for further control of the HC component reduced the HC+NO$_x$ to an average 7.0 g test^{-1} which is 53% below the limit for small cars. The addition of an oxidation catalyst in the exhaust gas stream also oxidises the carbon monoxide present so that the level is estimated to be about 50% lower than without the catalyst.

In the medium size car range (1.4 to 2.0 litres) the Luxembourg proposed limits are 30 g CO and 8 g HC+NO$_x$ per test. The GTL evidence shows that these limits are estimated to be exceeded by current lean-burn engine technology; however the addition of an oxidation catalyst, on average, resulted in CO emissions that were 50% below the limit and combined HC and NO$_x$ emissions that were just below the limit. (It is emphasised that the *GTL* figures were not intended to represent specific *Rover Group* vehicles, current or future.)

Table 2.3 also reports data obtained on experimental vehicles. Thus Menne and Stojek studied the lean-burn capability of a modified 2 litre *Ford* Sierra. Without an oxidation catalyst the CO and HC+NO$_x$ emissions were 12 and 7 g test^{-1} approximately; also, the NO$_x$ emissions were reduced by about 60% as compared with the conventional engine build. With the addition of an oxidation catalyst the CO and HC+NO$_x$ emissions were 3.5 and 3.5 g test^{-1}, respectively, which were below the requirements for large engines (25 g CO and 6.5 g HC+NO$_x$ with a maximum of 3.5 g NO$_x$). The fuel consumption on the ECE R15 cycle was reduced by 15% for this engine in the lean-burn configuration.

Experiments with a 2 litre Porsche and a 1.6 Passat showed that the Luxembourg levels are complied with by the application of lean-burn technology. However, there may be a fuel economy penalty if a minimum emissions tune is compared with a minimum fuel consumption tune. Thus the TNO data on the Porsche and Passat show that the minimum emissions tune may result in about 11% more fuel consumed on the ECE R15 cycle.

Figure 2.42 shows the NO$_x$ emissions range of the lean-burn ECE R15 results compared with the emissions envelope of the 20 car survey. It can be seen that the lean-burn concept produces very low emissions as compared with conventional engines. However, it should be noted that Figure 2.42 compares results from lean-burn experimental engines in vehicles, tested in accord with ECE R15-04 prescriptions to precise laboratory conditions, with results obtained on the road under normal in-use operating conditions with a range of in-service vehicles.

In an experiment designed to compare the emissions of low polluting modern cars with the conventional spectrum of vehicles in the UK fleet, WSL tested three lean-burn cars (A/F ratio 17–18:1), which had been built to the requirements of ECE R15-04, and one 3-way catalyst vehicle on the road over a wide range of operating conditions. The data for urban, suburban, rural and motorway drives were obtained under normal driving conditions in the traffic stream. Above the UK speed limit of 113 km h^{-1} (70 mph) the data were obtained in transient drives on a high speed test track. Figures 2.43, 2.44 and 2.45 show the emissions of CO, HC and NO$_x$ compared with the 20 car survey. It should be noted that the speeds are *average* speeds over routes considered to be typical of UK driving situations.

In the case of CO emissions (Figure 2.43) Cars A, B and C produced relatively low emissions up to about 100 km h^{-1} as compared with the 20 car survey. As the average speed increased above 100 km h^{-1} the emissions increased rapidly and exceeded the low speed (about 20 km h^{-1}) urban emissions values which were obtained under congested traffic conditions. These observations were consistent with the lean-burn operation of engines; thus, when maximum power is required richer A/F ratios obtain and CO emissions are increased. The 3-way catalyst car produced very low emissions of CO at average speeds up to about 60 km h^{-1}. At higher speeds the emissions increased, somewhat, and at speeds near to the maximum of the vehicle there was considerable dispersion of the data (not shown).

Figure 2.44 shows that the hydrocarbon emissions from the three lean-burn cars were low as compared with the emissions envelope of the 20 car study. It is of interest to note that the emissions of HC from Car A were below the lower boundary of the 20 cars envelope up to an average speed of about 95 km h^{-1}. The 3-way catalyst car was a very low emitter of HC throughout the speed range studied.

The emissions of NO$_x$ from the three cars equipped with lean-burn engines occurred in a band about the median of the 20 cars envelope. At high speeds, Figure 2.45 shows that the NO$_x$ emissions did not continue to rise but fell at maximum speed, which is consistent with the fuel enrichment requirements for maximum power which also produced the very high CO emissions shown in Figure 2.43.

COSTS

The cost of a lean-burn control system depends largely on the size-bracket of engine in the vehicle to meet the respective emission control figures. It is expected that by the time the Luxembourg Agreement Regulations are implemented, most cars will have electronically controlled ignition and fuel systems. The extra

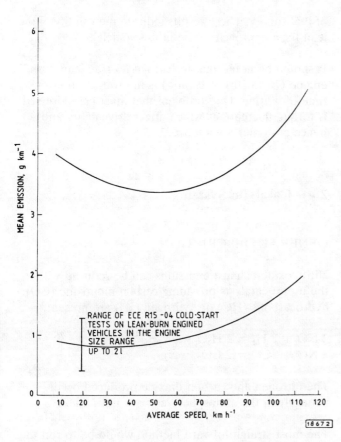

Figure 2.42 Variation of NO$_x$ Emissions with Average Speed from Lean-Burn Engined Cars Compared with the WSL '20 Car Survey' Emissions Envelope

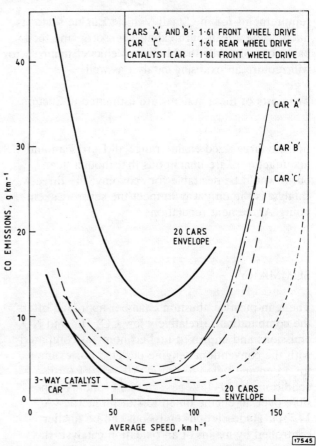

Figure 2.43 Variation of Carbon Monoxide Emissions with Average Speed from Lean Burn Engined Cars Built to ECE R15-04

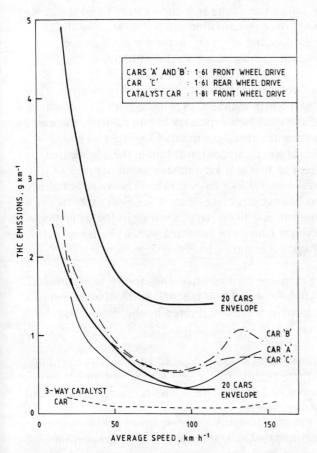

Figure 2.44 Variation of Total Hydrocarbons Emissions with Average Speed from Lean Burn Engined Cars Built to ECE R15-04

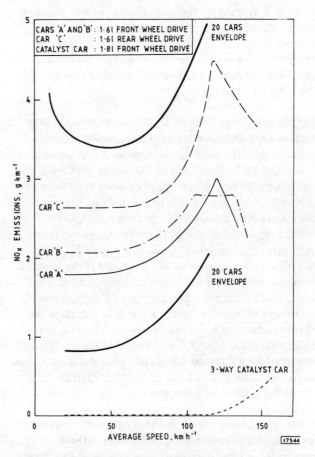

Figure 2.45 Variation of Nitrogen Oxides Emissions with Average Speed from Lean Burn Engined Cars Built to ECE R15-04

requirements for the Small Vehicle Engine sector is likely to be a lean-burn exhaust sensor giving feed-back control, while the Medium Vehicle Engine sector will require an oxidising catalyst as well.

The costs of these systems are estimated in Section 2.5.

For the large sized engine range all European and US manufacturers are unanimous that though 'lean-burn' would be desirable for economy, the three-way catalyst is the only way to meet the severe Luxembourg Agreement regulations.

SUMMARY

The lean-burn combustion chamber approach offers the combination of relatively low CO, HC and NO_x emissions and improved fuel economy as compared with the conventional engine designs. For example, for an engine tuned to run at A/F = 20:1 (cruise conditions) a 10–15% gain in fuel economy is possible over an engine tuned to stoichiometric (A/F = 14.7:1). Hydrocarbons emissions can be further controlled by means of an oxidation catalyst; this approach enables manufacturers to meet the proposed Luxembourg standards for HC+NO_x in the medium car size range (1.4–2.0 litres) without resort to three way catalyst systems. Furthermore, simple or sophisticated engine control systems may be applied to meet emissions legislation without fuel economy penalties. It should be noted, however, that if an oxidation catalyst is used, then unleaded fuel (RON 95) must be used.

The current generation of lean burn engines in production cars probably are more accurately described as fast burn. The A/F ratios employed are about 17:1 (*VW*) and 18:1 (*Ford, Austin Rover*) at part load conditions. Furthermore, these engines are calibrated to comply with the ECE R15-04 emissions regulations. The results of research into vehicles with engines designed to run in the true lean-burn range (A/F ratio 18:1 or more) show that in the ECE R15-04 test very low emissions are obtained if the concept is combined with an oxidation catalyst and, in some cases, EGR. For example a *Porsche* 2 litre engined car in 'emissions tune' with an oxidation catalyst and EGR produced 8.6 g CO and 5.4 g HC+NO_x per ECE test; a 1.6 litre Passat with an oxidation catalyst produced 16 g CO and 3.8 g HC+NO_x per test, whereas the proposed Luxembourg standard is 30 g CO and 8 g HC+NO_x per test.

In the case of current production cars with 'fast burn' engines, on-the-road tests showed that carbon monoxide and hydrocarbons emissions were relatively low, as compared with an earlier survey of in-service cars, whereas NO_x emissions were in the mid range

of that survey. These results indicate the validity of the lean burn approach to the in-use vehicle.

It should be noted that in Europe no true lean burn engine (18 to 20:1 A/F ratio) is in production at the time of writing. One problem that must be addressed is torque fluctuations which affect driveability and hence customer acceptance.

2.4 Catalytic Systems

GENERAL DESCRIPTION

Nitric oxide exhaust emissions can be reduced with the use of catalysts providing carbon monoxide (CO), hydrocarbons (HC) or hydrogen (H_2) are present.

$$2\ NO + 2\ H_2 = 2\ H_2O + N_2$$
$$2\ NO + 2CO = 2\ CO_2 + N_2$$

The nitrogen and carbon dioxide produced occur naturally in the atmosphere and are not toxic.

The most straightforward method would be to run the engines slightly rich of the stoichiometric A/F ratio and pass the exhaust gases over a reduction catalyst. Air can then be added from a simple air pump to the exhaust from the reduction catalyst and the gases passed over an oxidation catalyst to remove the excess CO and H_2. This method is not however favoured because it is wasteful of fuel.

The preferred method is the three-way catalyst (TWC) which manages to achieve both the reduction and the oxidation processes in one catalyst, thereby lowering the three pollutants CO, HC and NO_x. This, however, necessitates tuning the engine fuel system so that it is kept at stoichiometric plus or minus about 0.1 of an A/F ratio. This is achieved by feed-back control (see Section 2.1 Fuel Systems) in conjunction with an oxygen sensor in the exhaust gas. A typical European car fitted with a three-way catalyst system is shown in Figure 2.46.

The emission control situation is shown in Figure 2.47. It will be noted that if the A/F fluctuates between the narrow window indicated by the dotted lines both reduction and oxidation processes are optimised.

Catalysts are poisoned by lead and therefore require lead free petrol. This means a reduction in octane number and an increase in fuel consumption of about 5% (see Section 4.1).

Johnson Matthey (*JM*) supply a large proportion of the catalysts used on passenger vehicles principally in the USA. Their reduction catalyst uses platinum-rhodium deposited on a refractory washcoat lying on

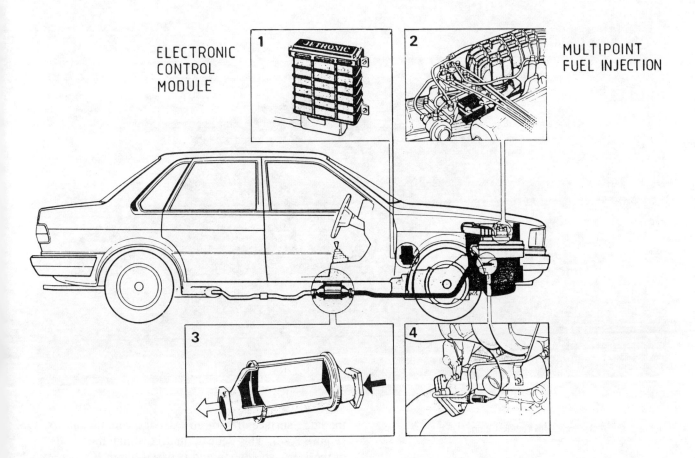

Figure 2.46 Three-Way Catalyst Car: European Concept (VW)

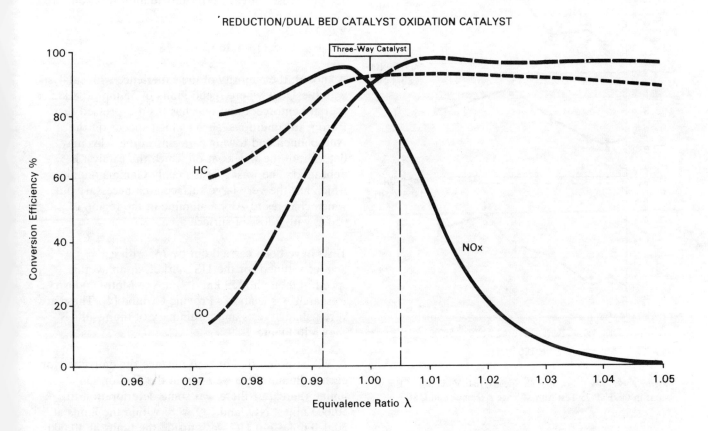

Figure 2.47 Selectivity of Rhodium/Platinum Alloy Catalysts

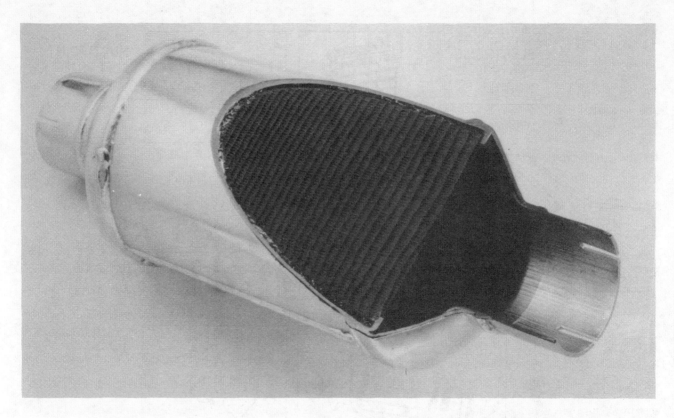

Figure 2.48 A Sectioned JM Catalyst Unit

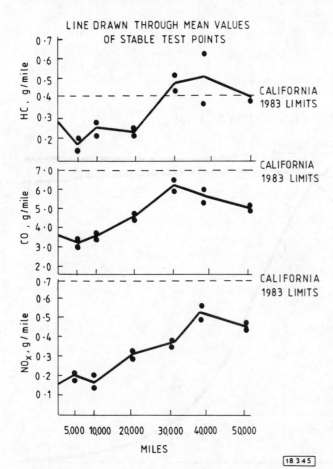

Figure 2.49 VW Scirocco High Speed Trial: Variation of Emissions in the FTP-75 Test with Mileage Accumulation (JM)

the large surface area of an extruded ceramic matrix (Figure 2.48). This delicate matrix is manufactured in the USA and Japan and is based on an ICI patent. A specialised mounting system is required.

Oxidation catalysts which may be used in lean burn engines usually employ a platinum palladium mix but are otherwise similar. Typical oxidation reactions are:

$$2\,CO + O_2 \rightarrow 2\,CO_2$$
$$2\,C_8H_{18} + 25\,O_2 \rightarrow 16\,CO_2 + 18\,H_2O$$

As noted, the majority of the experience with catalysts is in the USA where speed limits of 55 mph (88 km h^{-1}) are enforced, and fear has been expressed that European conditions, such as high speeds on the Autobahnen and towing caravans in the Alps may deteriorate the washcoat on which the catalyst is deposited. The washcoat is mainly Gamma alumina which has the very large surface area necessary but which changes to Alpha alumina in the region of 950°C with a loss of surface area.

Tests have been carried out by *JM* with a *VW Scirocco*, tuned for the US market, at an average speed of 99 mph (159 km h^{-1}) at the Motor Industry Research Association's Proving Ground [4]. These are very valuable tests and a summary of the results is shown in Figure 2.49.

It will be noted that at zero mileage the emissions for each pollutant were well within the Californian limits; thereafter there was some deterioration to 50,000 miles. NO$_x$ and CO were within the limits at 50,000 miles but HC was outside the limits at 40,000 miles. It should be further noted that Californian regulations require primarily 0.4 gm/mile for NO_x but 0.7 gm is allowed (until 1989) but recall may be implemented up to 75,000 miles.

The engine for this vehicle was of 1.8 litre capacity and was fitted with a *Bosch* 'K' Jetronic Fuel Injection system with a closed loop oxygen sensor.

It will be recognised that this was a severe test and the catalyst conversion efficiencies, at the end of the test, of 73, 72 and 69% for HC, NO_x and CO respectively were very creditable.

Catalyst temperatures ranged between 850–950°C indicating the washcoat and catalyst were well stabilised. The catalyst was Platinum/Rhodium in the ratio 5:1.

EMISSIONS AND FUEL CONSUMPTIONS

The emissions of some European three-way catalyst cars when tested in accord with the 1975 US Federal test procedure are given in Table 2.4.

Table 2.4 shows that production European cars when fitted with 3-way catalyst systems and full engine management met the US Federal 1983 and Californian standards. In the case of the *Volkswagen* Scirocco which had been fitted with a 'European' specification *Johnson Matthey Chemicals* three-way catalyst, the US Federal CO standard was exceeded at the 50,000 mile test point; however, the Californian standards of 4.35 g CO, 0.25 g HC and 0.43 g NO_x per km on the US 1975 driving schedule (see Section 1.3) were not exceeded. Furthermore, the 50,000 miles had been driven at an average speed of 99 mph as described in Section 2.4 and even under such severe conditions the catalyst efficiency remained high at 69, 73 and 72% for CO, HC and NO_x respectively.

The emissions of some catalyst cars when tested in accord with the prescriptions of ECE R15-04 are shown in Table 2.5. For convenience and for comparison with the 'Luxembourg Agreement' limits the table is divided into three sections, viz. small cars of engine size less than 1.4 litres, medium size cars of engine size 1.4 to 2.0 litres and large cars of engine size greater than 2.0 litres. All the cars were fitted with three-way catalysts; however, it can be seen from the table that experiments were carried out with vehicles fitted with open loop three-way catalyst systems, whereas it is usual for three-way catalyst cars to be equipped with a closed loop fuel injection control system, in which an oxygen sensor in the exhaust gas stream detects the A/F ratio of the mixture. These open loop three-way catalyst systems have been applied to small cars where expensive fuel injection and advanced engine management systems would normally be excluded.

In the case of small cars (engine size less than 1.4 litres) Table 2.5 reports three sets of data. First, four cars (*Fiat* Uno, *VW* Golf, *Rover* 213 and *Peugeot* 205) which had been purchased by *Johnson Matthey Chemicals* as standard cars but which were tuned to run on 95 RON unleaded petrol. All four cars were conventionally fuelled by carburettors. Searles [5] reported that the only modifications made were to the exhaust systems to allow the insertion of the catalyst containers and, where necessary, the engines were tuned to the manufacturers' specifications. The three-way catalyst formulation used in these experiments was 7.5% Rh/Pt/40. The target cost of the catalytic converter was less than £50.

The results show that the emissions from the four cars were below the Stage 1 Luxembourg standards for *small* cars. Furthermore, the *Fiat* Uno and *Rover* 213 emissions were below the more stringent medium size standards of 30 g CO and 8 g HC+NO_x per test. The *Peugeot* 205 HC+NO_x emissions exceeded the 8 g test^{-1} requirement for *medium* size cars but was 43% below the requirement for small cars. In the case of the *VW* Golf the emissions of HC+NO_x were 20% below the medium car standard but the CO emissions exceeded the 30 g test^{-1} requirement for

Table 2.4 Emissions of European Three-Way Catalyst Equipped Cars: US 1975 Procedure

Vehicle/ Manufacturer	No. of Vehicles	Pollutant Emissions, g km^{-1}		
		CO	HC	NO_x
Audi/Passat (M)	37	1.73	0.24	0.38
Audi/Passat (A)	28	2.02	0.18	0.53
Golf/Jetta (M)	11	2.02	0.23	0.40
Scirocco (new)	1	2.20	0.16	0.08
Scirocco (50 K miles)	1	3.09	0.25	0.29
Swiss investigation 'A'	1	1.30	0.16	0.09
Swiss investigation 'B'	1	1.22	0.21	0.14
US Federal 1983 standards, g mile^{-1}		3.4	0.41	1.0
US Federal 1983 standards, g km^{-1}		2.1	0.25	0.62
Californian 1983 standards, g km^{-1}		4.35	0.25	0.43

medium size cars by 12.7 g. It should be noted that the above discussion has made comparisons of emissions from three small engined cars of about the same size range. However, small engines are fitted to 'medium' size cars of larger inertia weights which may present greater engineering difficulties. Furthermore, although Stage 2 standards for small engined cars had not been negotiated at the time of this report it should be noted that van Beckhoven [6] of the Netherlands has proposed limits for small cars in the ranges 20 to 30 g CO and 6 to 7 g HC+NO$_x$ per test which are more stringent than the agreed limits for the medium size cars. Hence a comparison of the data from the *JMC* four car study is useful as a bench mark.

Manufacturers would have an engineering goal at the Type Approval (TA) stage which is 20% below the legal limit; in this regard it is interesting to note that the Uno, Golf and Rover 213 when fitted with open loop three-way catalysts had HC+NO$_x$ emissions that were 20% or more below the 8 g test^{-1} for medium size cars. A further consideration is that in production the Conformity of Production (CoP) standards are about 20% less stringent than at the TA stage. Thus the medium size HC+NO$_x$ CoP standard is 10 g test^{-1}.

The data presented by Searles indicated that the NO$_x$ conversion rates were about 25, 56, 65 and 61%, respectively, for the Peugeot 205, Uno, Golf and Rover 213. In the case of CO oxidation the lowest conversion rate was 15% for the Golf whereas the highest conversion rate was 67% for the Peugeot 205.

Van Beckhoven [6] studied data from the TNO research institute in the Netherlands, *Johnson Matthey* and Type Approvals. The experimental data were considered to represent 75% of the limit value for each group, i.e. he multiplied the results by 1.33. These data and *Porsche* data have been included in Table 2.5. It can be seen from the ranges of values reported that whereas all the vehicles would meet the Luxembourg Stage I standards, more stringent standards could not be met by all vehicles. Thus, for example, *Porsche* report 6–15 g test^{-1} of HC+NO$_x$ and thus would be in breach of an 8 g test^{-1} standard. However a controlled 3-way catalyst system using single point injection (TBI) gave 14–19 g CO and 2.5–6.0 g HC+NO$_x$ which is well below the medium size car standard.

The TUEV [7] reported very low emissions for a 1.3 litre *Opel* Kadett equipped with multipoint fuel injec-

Table 2.5 Emissions of European Three-Way Catalyst Cars: ECE R15-04 Procedure

Vehicle	Technology	Pollutant Emissions, g test^{-1}			
		CO	HC	NO$_x$	HC+NO$_x$
Small cars: less than 1.4 litres					
1.0 litre Fiat Uno	(carb) 3-way catalyst: open loop	9.8	1.4	2.7	4.1
1.3 litre VW Golf	(carb) 3-way catalyst: open loop	42.7	4.4	2.0	6.4
1.3 litre Rover 213	(carb) 3-way catalyst: open loop	27.5	3.8	1.4	5.2
1.3 litre Peugeot 205	(carb) 3-way catalyst: open loop	8.8	2.7	5.8	8.5
'R15-04' (TNO)	(carb) 3-way catalyst: open loop	18.0	–	–	6.8
'R15-04' (Porsche)	(carb) 3-way catalyst: open loop	12–30	–	–	6–15
'R15-04' (JMC)	(carb) 3-way catalyst: open loop	12–55	–	–	5.3–11.3
'R15-04': (Porsche)	(TBI) 3-way catalyst: closed loop	14–19	–	–	2.5–6.0
1.3 litre Opel Kadett	(MPI) 3-way catalyst: closed loop	3.8	1.4	2.0	3.4
Luxembourg Stage 1 standard: small cars		45	–	6	15
Medium size cars: 1.4 to 2.0 litres					
1.5 litre Fiat Uno	(MPI) 3-way catalyst: closed loop	14.4	1.3	0.59	1.89
1.8 litre Audi 100	(MPI) 3-way catalyst: closed loop	7.2	1.3	0.69	1.99
1.8 litre VW Passat‡	(MPI) 3-way catalyst: closed loop	19.0	2.5	0.46	3.00
1.8 litre VW Scirocco*	(MPI) 3-way catalyst: closed loop	15.4	1.1	0.39	1.49
2.0 litre D-B 190E†	(MPI) 3-way catalyst: closed loop	24.7	4.9	3.6	8.5
2.0 litre Ford Sierra	(MPI) 3-way catalyst: closed loop	6.4	1.0	2.0	3.0
2.0 litre Engine simulations (Ford UK)		9.8	2.7	0.56	3.3
Luxembourg standards: medium cars		30	–	–	8
Large cars: greater than 2.0 litres					
2.2 litre VW Passat	(MPI) 3-way catalyst: closed loop	8.5	1.7	0.92	2.6
2.3 litre Volvo 740	(MPI) 3-way catalyst: closed loop	15.3	1.8	2.7	4.5
2.7 litre BMW 325	(MPI) 3-way catalyst: closed loop	8.7	1.7	1.5	3.1
Luxembourg standards: large cars		25	–	6	15

Table 2.6 Comparison of the Fuel Consumption of Catalyst and non-Catalyst Cars (litres per 100 km)

	Hot Start ECE R15			90 km h^{-1}			120 km h^{-1}		
	Baseline	Catalyst	% Change	Baseline	Catalyst	% Change	Baseline	Catalyst	% Change
1.0 litre Fiat Uno	6.83	6.82	–	4.68	4.87	+4.0	6.21	6.38	+2.7
1.3 litre VW Golf	8.14	7.87	−3.3	5.16	5.12	–	6.43	6.63	+3.1
1.3 litre Rover 213	7.87	7.98	+1.4	4.96	5.02	+1.0	6.43	6.54	+1.7
1.4 litre Peugeot 205	7.51	7.85	+4.5	4.58	4.60	+1.0	5.81	5.82	–
1.8 litre VW Scirocco	12.2	12.6	+3.3	–	–	–	–	–	–
2.0 litre Sierra	–	–	+5	–	–	–	–	–	–

tion and oxygen sensor feed-back control. Thus the 3.8 g CO, 2 g NO$_x$ and 3.4 g HC+NO$_x$ per test were respectively 15, 33 and 23% of the *large* car standard.

In the medium size car range Table 2.5 reports three-way catalyst car data for engine sizes ranging from 1.5 to 2 litres. It can be seen that with the exception of the *D-B* 190E all cars readily met the medium size car limits and were below the more stringent limits for the large cars. TUEV reported a fault in the case of the *D-B* 190E which should not therefore be regarded as typical. The *Fiat* Uno, *Audi* 100, *VW* Passat and *D-B* 190E were all in-use cars. The *VW* Scirocco had been imported by *Johnson Matthey Chemicals* from California and fitted with European specification catalysts. The *Ford* Sierra data were determined by Menne [8] as part of a comparative study of emissions from three-way catalyst, standard and lean-burn concept engines. In the case of the *Ford* UK engine studies it can be seen that the projected emissions are not at variance with the car measurements.

Three sets of car data were determined by TUEV [7] for large cars. It can be seen that whereas the highest emitter was the *Volvo* 740 its emissions of CO, NO$_x$ and HC+NO$_x$ were 61, 45 and 30% respectively of the large car standard.

Table 2.6 reports the fuel consumptions of the four catalyst cars fitted with open loop three-way catalysts; also the *VW* Scirocco and 2 litre *Ford* Sierra. Although the engines of the four small cars were not re-calibrated for catalyst operation fuel consumption changes were observed from 3.3% less to 4.5% more fuel consumed.

The Scirocco and Sierra were fitted with fuel injection and full engine management. Also, in the case of the Scirocco the baseline condition was measured without the catalyst but with full engine management and Lambda control retained. The Sierra comparison was between the standard non-catalyst concept and full three-way catalyst with feedback Lambda control. In general it is accepted that up to 5% more fuel is consumed by the application of catalytic control devices at Lambda = 1 (A/F about 14.7:1).

Road Emissions Data

The *VW* Scirocco tests described above were extended to include on-the-road emissions tests on three vehicle 'builds'. These were (i) baseline in which the catalyst was replaced by an uncoated ceramic monolith canned in a similar manner, (ii) the 50,000 mile durability test catalyst, and (iii) a 200 h aged catalyst which was considered to have been exposed to modest thermal stress. The test procedure that was applied to all three conditions is shown in Table 2.7.

Table 2.7 Test Procedures as Applied to all Three Vehicle Conditions

Test Centre	Procedure
WSL	1 × cold start ECE R15-04
	5 × urban route 1
	5 × urban route 2
	4 × rural route
	5 × motorway drives (90 kph)
	5 × motorway drives (113 kph)
MIRA	5 × drives at 160 kph (5 circuits per test)
	5 × drives at wide open throttle (WOT) (5 circuits per test)
	1 × cold start ECE R15-04

The fuel used was 91 RON as required by the Californian calibration of the vehicle.

The results for the three conditions of the vehicle are presented in Figure 2.50 and compared with the emissions envelope of the 20 car survey described in Section 2.2. These results are those obtained on the public highway; at higher speeds on the test track the emissions data are given in Table 2.8 along with ECE R15 cold and hot start data. The average conversion efficiencies of the two catalysts are given in Table 2.9.

The major conclusions of the work were:

(1) In the baseline and 50,000 mile catalyst conditions the carbon monoxide emissions decreased as the trip average speed increased up to approximately 60 kph when the average emissions were respectively 6.3 and 2.1 g km^{-1}. Higher speeds up to the maximum (168 km/h) resulted in increasing emissions. The 200 h catalyst condition produced

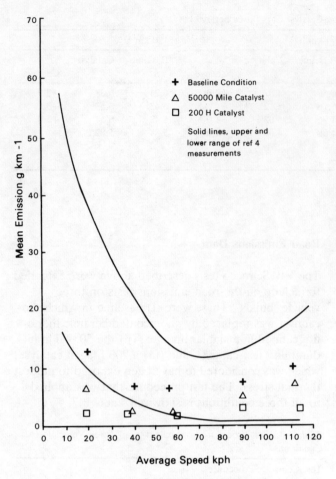

Variation of CO Emissions with Average Speed

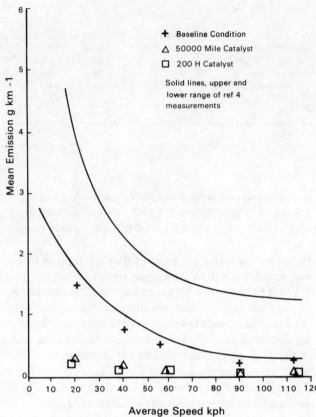

Variation of HC Emissions with Average Speed

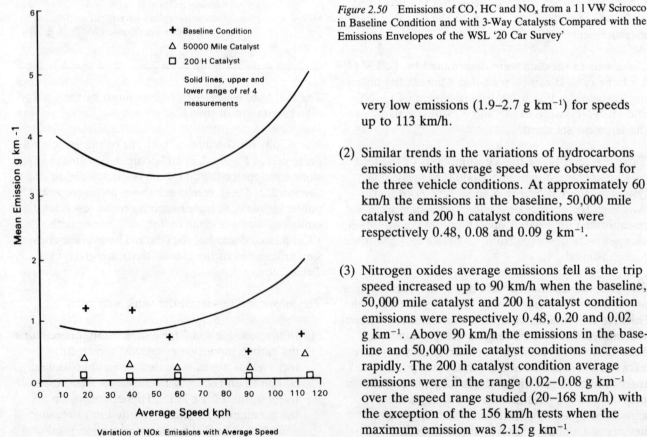

Variation of NOx Emissions with Average Speed

Figure 2.50 Emissions of CO, HC and NO$_x$ from a 1 l VW Scirocco in Baseline Condition and with 3-Way Catalysts Compared with the Emissions Envelopes of the WSL '20 Car Survey'

very low emissions (1.9–2.7 g km^{-1}) for speeds up to 113 km/h.

(2) Similar trends in the variations of hydrocarbons emissions with average speed were observed for the three vehicle conditions. At approximately 60 km/h the emissions in the baseline, 50,000 mile catalyst and 200 h catalyst conditions were respectively 0.48, 0.08 and 0.09 g km^{-1}.

(3) Nitrogen oxides average emissions fell as the trip speed increased up to 90 km/h when the baseline, 50,000 mile catalyst and 200 h catalyst condition emissions were respectively 0.48, 0.20 and 0.02 g km^{-1}. Above 90 km/h the emissions in the baseline and 50,000 mile catalyst conditions increased rapidly. The 200 h catalyst condition average emissions were in the range 0.02–0.08 g km^{-1} over the speed range studied (20–168 km/h) with the exception of the 156 km/h tests when the maximum emission was 2.15 g km^{-1}.

Table 2.8 Summary of Average VW Scirocco Emissions (g km^{-1})

	Baseline Condition					50,000 Mile Catalyst Condition					200 Hour Catalyst Condition				
	Average Speed, km/h	CO	HC	NO$_x$	Fuel consumption, l(100 km)$^{-1}$	Speed, km/h	CO	HC	NO$_x$	Fuel consumption, l(100 km)$^{-1}$	Speed, km/h	CO	HC	NO$_x$	Fuel consumption, l(100 km)$^{-1}$
ECE (cold)	19	10.4	1.03	1.22	13.9	19	4.88	0.29	0.29	15.1	19	1.03	0.13	0.13	12.2
ECE (hot)	19	9.11	1.25	1.02	12.2	19	2.53	0.16	0.19	12.6	–	–	–	–	–
Hitchin	20.4	12.69	1.51	1.19	12.5	18.9	6.39	0.27	0.38	14.3	18.5	1.87	0.22	0.08	12.2
Stevenage	40.4	6.66	0.78	1.16	7.2	38.6	2.31	0.16	0.26	7.3	38.2	1.81	0.14	0.06	8.2
Cross-country	55.5	6.34	0.54	0.73	6.7	58.5	2.06	0.09	0.22	6.1	59.4	1.59	0.10	0.07	6.4
90 km/h	89.6	7.10	0.22	0.48	6.3	89.8	4.75	0.05	0.20	6.5	89.5	2.92	0.08	0.02	6.7
113 km/h	112.4	9.77	0.28	0.76	7.9	113.1	6.99	0.07	0.44	8.2	114.8	2.66	0.07	0.02	8.7
158 km/h	154.8	14.82	0.42	4.69	12.3	155.1	18.82	0.29	2.55	11.0	156.5	12.01	0.17	0.49	13.2
W.O.T.	167.3	39.17	0.60	3.94	15.2	168.0	25.68*	0.35	2.19	13.1	167.9	22.63	0.37	0.06	14.5
ECE (cold)	19	10.86	1.17	1.25	14.5	19	3.31	0.39	0.24	11.3	19	2.67	0.21	0.04	11.9
											19	4.57	0.31	0.09	13.2

*Average of 6 results: single result of 67.17 g km^{-1} excluded as an outlier at the 95% confidence level

Table 2.9 Average Conversion Efficiencies of the Catalysts, per cent (calculations are based on average values for the three conditions reported in Table 2.8)

Test	500,000 Mile Durability Catalyst			200 Hour Catalyst		
	CO	HC	NO$_x$	CO	HC	NO$_x$
ECE (cold)	54	72	76	90	87	89
ECE (hot)	72	87	81	–	–	–
Hitchin	50	82	68	85	85	93
Stevenage	65	79	78	73	82	95
Cross-country	68	83	70	75	81	90
90 km/h	33	77	58	59	64	96
113 km/h	28	75	42	73	75	97
158 km/h	–	31	46	19	60	80
W.O.T.	34	42	44	42	38	98
ECE (cold)	70	67	81	75	82	97
ECE (cold)				58	74	93

(4) High speed drives in which there is a transient component may cause higher emissions of the three regulated pollutants from a 3-way catalyst system than steady-state drives at similar speeds.

(5) The average urban drive emissions of the 50,000 mile and 200 h catalyst conditions were significantly lower than the lowest CO, HC and NO$_x$ emissions measured in a survey of European specification cars (respectively 9.9, 1.8 and 0.8 g km^{-1}). For example, the 50,000 mile catalyst emissions of CO, HC and NO$_x$ were 6.4, 0.27 and 0.28 g km^{-1}.

COSTS

The cost of a total catalyst system is much influenced by the engine size as it relates to the Luxembourg Agreement figures and also to the engine size in relation to the weight of the vehicle which it powers, as this affects the concentration of the pollutants. The total systems are therefore more appropriately considered in Section 2.5. The cost of the catalyst boxed in its high temperature metal container, suitable for incorporation in the exhaust pipe, depends somewhat on its size as below.

3 Way Catalysts suitable for Engine Range
1.4 litre to 2 litre   c. £80

3 Way Catalyst suitable for above 2 litre   c. £100

SUMMARY

Three way catalyst systems with full engine management control are considered to be the most effective current technology to reduce gaseous pollutant emissions from spark ignition engines to very low levels. Carbon monoxide and HC emissions are controlled by oxidation to CO_2 and water whilst NO$_x$ is controlled by reduction in the presence of CO and H_2 to nitrogen, CO_2 and water. The system controls the three regulated pollutants efficiently only at an air/fuel ratio close to stoichiometric. It is important to note that all catalyst systems require unleaded fuel because lead and lead compounds rapidly poison the known catalyst formulations.

Generally, three way catalyst systems for large cars (>2 litres) are fuelled by multipoint fuel injection systems under electronic control in a closed loop configuration which employs an oxygen sensor in the exhaust gas stream to provide a feed-back signal to the electronic control module. However, single point fuel injection has been used as a less expensive option.

The conversion efficiencies of three-way catalyst systems are about 80%, 90% and 95% for CO, HC and NO$_x$ respectively. In the case of the proposed Luxembourg limits for large cars (25 g CO, 6.5 g HC+NO$_x$ and 3.5 g NO$_x$ per test) all manufacturers agree that the three-way catalyst system is the most appropriate technology. It should be noted that this technology results in a reduction in fuel economy of about 5% compared with lean-burn engines.

However, it is important to note that the improved engine management systems and air/fuel ratio control can achieve fuel economies similar to those of a 'baseline' vehicle without catalyst and baseline calibration. Furthermore, catalyst deactivation can occur if the vehicle is fuelled with leaded petrol. There are steps which can be taken to ameliorate this problem although it would be difficult to overcome completely.

2.5 Application of Control Technologies

As will be seen from Section 1.2 more severe European legislation is proposed and in the first phase the proposed emission limits are shown again for convenience in Table 2.10.

Table 2.10 The Luxembourg Agreement Proposed Limits for Light Duty Vehicles

Cylinder Capacity (litres)	Mass of Carbon Monoxide (g/test)	Combined Mass of Hydrocarbons and Nitrogen Oxides (g/test)	Mass of Nitrogen Oxides (g/test)
greater than 2 l	25	6.5	3.5
1.4 l to 2 l	30	8	–
less than 1.4 l*	45	15	6

*Stage I limits only; Stage II still to be agreed

It will be seen that the reductions compared with the previous regulation (see Section 1.3) are less severe for the small vehicle and become more severe for vehicles with engines between 1.4 l and 2 l. With vehicles having engines above 2 l they are very severe, indeed of similar severity to the US regulations.

Successive amendments to the original Regulation 15-00 have become increasingly stringent. For carbon monoxide the emission limit has decreased by 50% from a value of 117 g/test (for cars of reference weight 850–1020 kg) in the original ECE Regulation 15, to a value of 58 g/test in Amendment 04. It is less straightforward to compare the individual hydrocarbons (HC) and NO_x limits in 15-00 and 15-04 as only a combined HC and NO_x limit is quoted in 15-04, and no NO_x limit was quoted in 15-00. From 15-00 to 15-03, the HC limit decreased by 25% from 8.7 g/test to 6.5 g/test (for 850–1020 kg cars). The $HC+NO_x$ limit decreased from 15-02 to 15-04 from 27.0 g/test to 19.0 g/test, a decrease of 30%.

In terms of the limit values given in Tables 1 and 2, the 'Luxembourg Agreement' represents a substantial decrease over Amendment 15-04. Comparisons with the Luxembourg Agreement are not straightforward given that 15-04 is framed in terms of vehicle weight and the Luxembourg limits are given in terms of engine size. However, to a good approximation it can be assumed, for the UK at least, that an average of the 850–1020 kg and 1020–1250 kg limits gives the 'UK fleet average' limit. Similarly using figures of 8.3%, 38.3% and 52% for the proportions of 2 litre, 1.4–2 litre and 1.4 litre cars the fleet-average 'Luxembourg Limits' can be derived. (These figures are SMMT data for 1985; 1.3% of vehicles are classified as 'unknown'.) On this basis the fleet average CO limits are 62.5 g/test for 15-04 and 37.0 g/test for Luxembourg, a decrease of 41%. For $HC+NO_x$ the corresponding figures are 19.8 g/test and 11.4 g/test, a decrease of 42%.

The Luxembourg limits for small cars are less stringent than those for the medium or large category, although Stage II limits for small cars are still to be agreed. It is possible therefore that, given the fraction of small cars in the UK fleet (52% in 1985), the greater part of vehicle emissions of CO and $HC+NO_x$ could arise from this category of vehicle. Some caution is necessary however since emission limits may not reflect accurately the actual emissions in service. Measurements will be necessary in the future to address this point.

The differences in the proposed regulations with respect to size have resulted in manufacturers addressing themselves to each size for the preferred solutions. These will be considered in turn.

VEHICLES WITH SMALL ENGINES: LESS THAN 1.4 LITRES

As is evident from the previous section it is quite possible to meet the Luxembourg proposed regulations by lean-burn without catalysts. This would obviate the necessity to use unleaded fuel; however it should be noted that it is the intention of the UK government and the European Commission to encourage the use of unleaded petrol. The second phase of the small cars limits (1992/93) *may* be as stringent as the medium car (1.4–2.0 l) limits in which case simple oxidation catalysts would probably be necessary and unleaded fuel a necessity. Some EGR may be necessary and precision in fuel metering and ignition timing will be essential.

Much experiment and development will be necessary to give acceptable driveability. For good acceleration it is necessary to change from an A/F ratio of about 20:1 to 15:1 almost instantaneously. Without multi-point injection this is very difficult because of the delay between the carburettor producing the new condition, and the engine receiving it and settling to the new condition, while at the same time it is accelerating in speed. This tends to give delay in vehicle response and unsteady acceleration.

The above factors will encourage the use of electronic control of both fuel and ignition. Single point injection may be preferred only because of the cost to the customer.

MEDIUM SIZED ENGINED VEHICLES: 1.4 TO 2.0 LITRES

VW point out that in Germany catalysts are well accepted by the public and that it is easier to meet production dates with this system rather than the very lean-burn approach necessary in this vehicle range. They are satisfied that the 3-way catalyst is reliable even under autobahn driving conditions.

Fiat's approach in this range is to make use of their experience in tuning their cars to meet the US 1983 regulations which are required in Austria, Sweden, Switzerland, and their choice will probably lie between the 3-way catalyst with a Lambda sensor, or lean-burn plus an oxidation catalyst and EGR. Fuel injection would be essential.

Ford (Europe) intend to continue their policy of the development of HCLB engines in this medium engine size range. It was also stated that the solutions below and above this engine range may also be applicable. Thus detailed attention to charge induction, lean burn (18:1) and engine management with TBI may be appropriate at the lower end of the engine size range. Whereas some solutions at the higher end of the range *may* include three-catalyst systems and full engine management.

LARGE CARS

Cars with engines above 2 litres in capacity are in comparatively small numbers and the regulations are more severe. As a consequence all manufacturers propose to start off with 3-way catalysts with Lambda sensors feeding back into multipoint fuel injection systems to control the A/F ratio to stoichiometric. This is not an economic solution as far as fuel consumption is concerned, but is likely to be the only way to meet proposed legislation in the limited time available. Programmed ignition will be universal on these more expensive vehicles.

In the US 3-way catalyst systems are widely used; the popular cars use single point injection and the upmarket multipoint; programmed ignition is universal.

GM are studying lean-burn approaches but do not consider them a likely contender in the near future because fuel in the USA is inexpensive and fuel costs do not worry the customer. *Ford* on the other hand consider that the less severe NO_x requirement for the Federal States in comparison with California (1.4 g/mile compared with 0.4 g/mile) makes the lean-burn approach a desirable target for the Federal States. The Proco stratified charge engine (see Section 5.1) was one approach in which a lot of work has been performed. They consider that the lean-burn conventional engine is well worth pursuing and in this connection they favour a moderate 17:1 A/F ratio plus oxidation catalyst plus EGR. *Ricardo* are of the opinion that the 3-way catalyst approach is the only viable one for these large engines.

COMPARISON OF EMISSIONS

The assessment of competing technologies to reduce pollutant emissions is not an easy task. Manufacturers state that difficulties in such comparisons arise due to the differing calibrations of the engine and control systems to meet specific design targets including legislation. However, in an attempt to compare emissions from a 2 litre engine, *Ford* provided data calculated from engine dynamometer steady state mapping points. Table 2.11 summarises the Ford data.

In the case of CO emissions, the engine when calibrated to the stoichiometric A/F ratio produced 19.2 g test^{-1} in the baseline condition. It was calculated that with TWC these emissions would be reduced by 49%. Similarly $HC+NO_x$ emissions were reduced by approximately 76%; the NO_x content of the combined $HC+NO_x$ was reduced by 91%.

Table 2.11 Ford 2 l Engine (EFI) Dynamometer Based Emissions Calculated for Simulated ECE (Hot) Tests

Engine Calibration	Simulated and Calculated Hot ECE Emissions, g test^{-1}			
	CO	HC	NO_x	$HC+NO_x$
Stoich. no catalyst (A/F 14.6:1)	19.2	7.21	6.27	13.48
Stoich. + TWC (calculated)	9.81	2.72	0.56	3.28
Notional lean-burn (A/F 18:1)	16.01	8.03	2.53	10.56
Notional lean-burn + oxidation catalyst (calculated)	9.58	2.85	2.53	5.38
Luxembourg standard (cold start)	30	–	–	8

Notes (i) The engine was *not* developed to utilise HCLB. The data were taken at the leanest sustainable A/F ratio to provide a comparison as requested by the Fellowship of Engineering.
(ii) Although these are hot test data, this engine in a vehicle equipped with a full TWC and engine management system has produced lower results in the full cold test procedure.

The notional lean-burn calibration in the baseline condition showed a reduction in CO and NO_x emissions as compared with the baseline stoichiometric A/F ratio conditions. Thus CO was reduced by about 17% and NO_x emissions by about 60%. However, it is important to note that in practice an engine would not be calibrated to stoichiometric fuelling unless a three-way catalyst system was included in the overall emissions control package.

The addition of an oxidation catalyst to the lean fuelling calibration (A/F 18:1) produced a calculated 40% reduction in CO emissions and a 65% reduction in the HC emissions, whereas the $HC+NO_x$

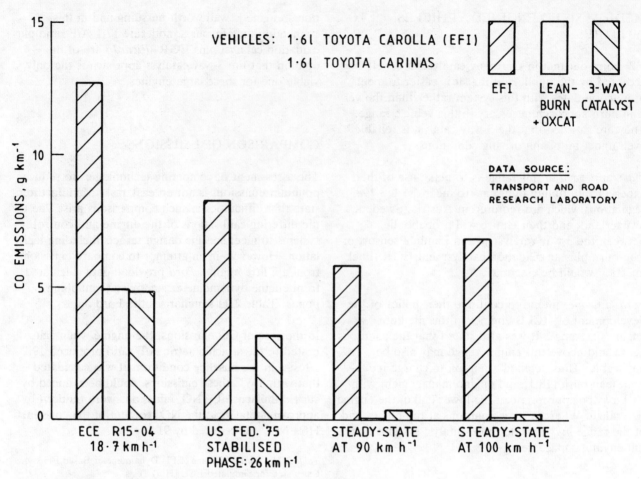

Figure 2.51 Comparison of CO Emissions from 1.6 l Cars: 3 Technologies

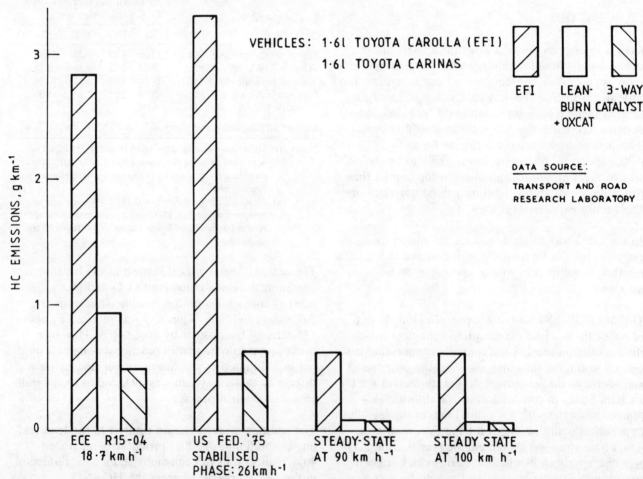

Figure 2.52 Comparison of HC Emissions from 1.6 l Cars: 3 Technologies

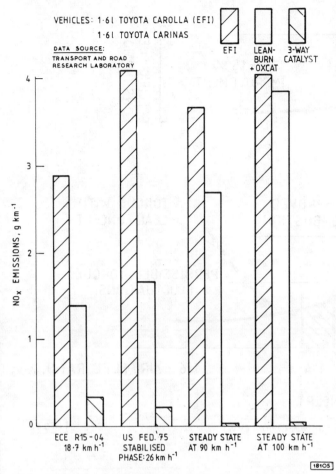

Figure 2.53 Comparison of NO$_x$ Emissions from 1.6 l Cars: 3 Technologies

that the highest emissions of the three regulated pollutants were emitted by the European tuned 1.6 litre EFI Corollas.

Figure 2.51 shows that in the ECE R15 test the lean-burn + catalyst build produced CO emissions that were approximately the same as the TWC build and about 60% less than the standard European configuration. In the stabilised phase of the US Federal 1975 drive the CO emissions from the lean-burn and three-way catalyst cars were 68 and 61% respectively lower than the standard build. However, under steady-state drive conditions at 90 and 100 km h^{-1} the CO emissions reduction of the advanced technology cars compared with the standard build were 95 to 97%.

Figure 2.52 shows that the advanced technology cars produced low HC emissions as compared with the 'European' car build. Thus in the cold start ECE R15 test the lean-burn + oxidation catalyst system emitted 67% less hydrocarbons than the standard European car. The three-way catalyst car emitted 83% less HC. At the higher speeds and under hot start conditions of the US Federal stabilised drive and steady-state speeds of 90 and 100 km h^{-1}, the HC emissions from the advanced technology cars were 81 and 92% lower than the standard build.

In the case of NO$_x$ emissions (Figure 2.53), the lean-burn + oxidation catalyst cars produced 51%, 59%, 27% and 5% less emissions respectively over the cold start ECE R15 cycle, stabilised US Federal drive and steady-state drives at 90 and 100 km h^{-1}. In contrast however, the three-way catalyst cars produced respectively 88%, 95%, 99% and 88% less emissions over the same series of tests.

The lean-burn engine NO$_x$ emissions results show that under relatively low average speed driving cycle conditions the engine produced about 50–60% less emissions than its equivalent 'European' calibration. However, at higher speeds NO$_x$ control was substantially reduced such that at 100 km h^{-1} the difference between the lean-burn and standard cars was only 5%. The Toyota lean-burn engine and its control system are probably tuned to the relatively low speed Japanese cycle, hence fuel enrichment at higher speed/power demands may be required. Kimbara et al. [9] described the Toyota lean-burn system in which the electronic control unit is programmed to calculate the required A/F ratio from the engine speed and intake manifold depression. Whereas idling requires a relatively rich mixture of A/F ratio, 15.5:1, normal acceleration and cruising requires respectively 22.5 and 21.5:1. However, at high speeds the A/F ratio is enriched to about 16:1, and under wide open throttle conditions the ratio is 12 to 13:1.

The effect of fuel enrichment can be seen in Figure 2.54 which is taken from various Toyota publi-

reduction was only 49% because the oxidation catalyst would not affect the NO$_x$ emissions.

The *Transport and Road Research Laboratory* (TRRL) has an on-going programme to compare emissions from Toyota 1.6 litre engined cars in three control technology configurations, namely, (i) European (ECE R15-04), (ii) full lean-burn with oxygen sensor (Lambda control) and oxidation catalyst and (iii) three-way catalyst with full engine management. The fuelling strategy in the three technologies represented is multipoint fuel injection. The lean-burn engines were stated to be fitted with an oxidation catalyst but there is some evidence in the Toyota manual that the catalyst is identical to the catalyst used in the three-way catalyst build (same Part No).

The vehicles are operated normally over a wide range of operating conditions. At about 3000 mile intervals the emissions are checked; at about 12,000 mile intervals, at which the major services become due, the emissions are measured before and after service.

Figures 2.51, 2.52 and 2.53 compare the emissions of CO, HC and NO$_x$ from the three vehicle groups when tested in accord with the ECE R15-04 procedure, US Federal 1975 cycle stabilised drive and steady-state tests at 90 and 100 km h^{-1}. It can be seen

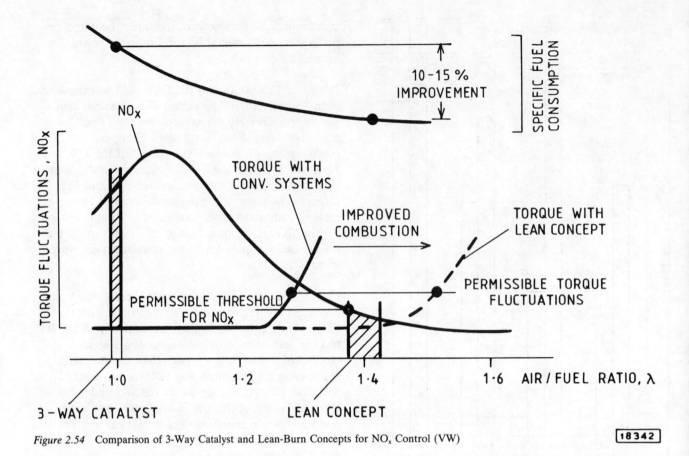

Figure 2.54 Comparison of 3-Way Catalyst and Lean-Burn Concepts for NO$_x$ Control (VW)

cations. It can be seen that when the A/F ratio is enriched from the lean-burn sector higher NO$_x$ emissions occur.

In the experiments described, the NO$_x$ control of the lean burn engine was successively reduced as the speed was increased. It should be noted that the steady-state speeds of 90 and 100 km h^{-1} would not be normally encountered on the road. Transient driving conditions obtain in the real situation and hence excursions into fuel rich conditions may occur. In this case it is reasonable to expect some NO$_x$ reduction to occur since the Pd/Pt oxidation catalyst of the lean burn cars will reduce NO$_x$ when rich A/F ratios obtain.

In the Federal Republic of Germany *TUEV* Koeln and *RWTUEV* Essen [2] have measured emissions from 37 light-duty vehicles of engine capacity ranging from 1.04 to 4.97 litres. The spark ignition engine group comprised 17 vehicles fitted with controlled 3-way catalyst systems, 2 vehicles fitted with uncontrolled 3-way catalyst systems, 3 vehicles equipped with pulsair systems (to increase CO and HC oxidation), 10 standard ECE R15-04 and 5 diesel engined vehicles.

The schedule of emissions testing included cold start ECE R15-04 tests followed by rural and highway driving simulations in accord with the proposals of the Committee of Common Market Constructors (CCMC), the UK and the FRG. Table 2.12 shows the average and maximum speeds of these candidate additional driving cycles to the ECE R15 procedure.

The CO, HC and NO$_x$ oxides emissions measured in accord with the prescriptions of ECE R15-04 are presented in Figures 2.55 to 2.57. It can be seen that the highest emissions of the three pollutants were from the ECE R15-04 vehicles, the diesel vehicles were low emitters of CO and HC and, generally, lower emitters of NO$_x$ than the standard (R15-04) cars. In the case of controlled 3-way catalyst cars the emissions of all three gaseous pollutants were low. However, the efficiency of the DB 190 E (Car No. 1) emission control system was considered to be too low.

Table 2.12 Average Maximum Speeds of the Rural and Highway Candidate Cycles

Proposal	Speed, km h^{-1}			
	Rural		Highway	
	Average	Maximum	Average	Maximum
CCMC	61	80	83	100
UK	56	70	85	100
FRG	59	?	100	11

In the ECE R15 cycles the emissions of pollutants varied widely in each control concept; for example, Figure 2.55 shows that the lowest CO emissions (vehicles 10 and 13) were about 3.8 g test^{-1} and the highest emissions were 23–25 g test^{-1} (vehicles 1 and 9) which were within the 'Luxembourg' agreement

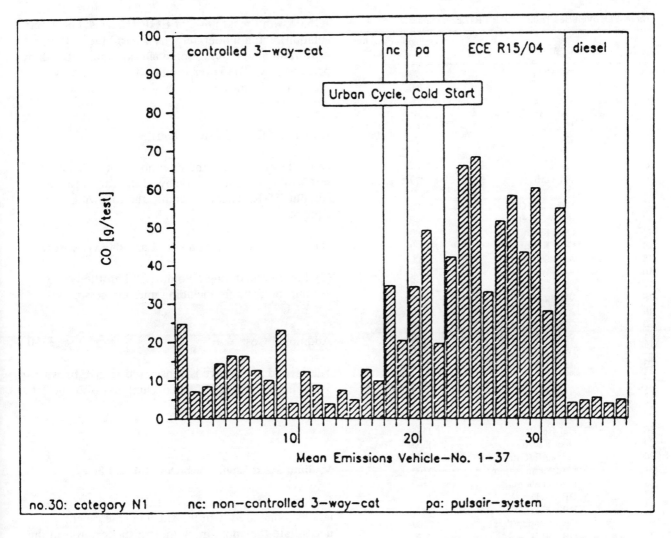

Figure 2.55 Comparison of CO Emissions in the ECE R15-04 Test from 37 Light Duty Vehicles Grouped According to Technology Class (TUEV Koeln & Essen)

standards for large cars. In the case of HC emissions (Figure 2.56) the lowest emissions were measured from the diesel engined cars and the highest emissions were from the non-catalyst R15-04 cars; the latter were however within the expected range for current conventional cars (see Table 2.13) with the exception of Car No. 32. Two 3-way catalyst cars (Nos. 1 and 9) exceeded the stringent 'Luxembourg' standards for large cars. The NO_x emissions from the different technology groups of vehicles in the ECE R15 test followed a similar pattern qualitatively. Thus the controlled 3-way catalyst cars and diesel cars were the lowest NO_x emitters (Figure 2.57) and the highest were the current R15-04 cars. Two cars equipped with controlled 3-way catalyst systems and two of the five diesel engined cars exceeded the 'Luxembourg' agreement for large cars NO_x emissions (3.5 g test^{-1}). Figure 2.58 shows the HC+NO_x emissions for these cars in the cold start test grouped according to technology class; it can be seen that the plots reflect the qualitative observations made with respect to the component pollutants. The calculated fuel consumptions are shown in histogram form in Figure 2.59 and should be considered in conjunction with the vehicle identities given in Table 2.13.

The emissions measured in the high speed tests (rural land highway drives combined) are given in Figures 2.60, 2.61 and 2.62 and expressed as g km^{-1}. The highest emissions of the three regulated pollutants were from the conventional ECE R15-04 cars; thus CO, HC and NO_x emissions of up to 10.7, 1.25 and 5.4 g km^{-1} were recorded. The lowest CO, HC and NO_x emissions were recorded from catalyst cars. Diesel NO_x emissions were also very low as compared with the R15-04 cars, thus the maximum NO_x emissions (from diesel car No. 35) were about 1 g km^{-1} and the maximum emission from a conventional gasoline car was 5.4 g km^{-1} (both measured in the FRG cycle).

In summary the data presented in Figures 2.55 to 2.62 (excluding 2.59) show the low emissions levels of the regulated pollutants that can be achieved by in-use catalyst equipped cars and diesel engined cars as compared with current conventional European specification cars. The TRRL data also show the potential for lean-burn engined cars which have the further advantage of reduced fuel consumption as compared with controlled three-way catalyst cars.

Table 2.13 Vehicles Tested in the FRG Programme

Vehicle No.	Vehicle Type	Engine capacity, litres
1*	DB 190 E	1977
2	Audi 100	1760
3	VW Passat	2200
4	Fiat Uno 75ie Cat.	1465
5**	VW Passat	1760
6	924 S	2479
7	928	4957
8	911	3164
9	944 Turbo	2479
10	DB 500 SE	4973
11	Volvo 740 GLE	2300
12	BMW 325 eta	2700
13	Opel Kadett	1300
14	Opel Kadett	1300
15	Audi 80 GTE	1800
16	Audi 100	2200
17	Ford Escort	1600
18	VW Polo	1035
19	Audi 90	2000
20	Citroen BX 16TRI	1569
21	Volvo 360 GLT	1954
22	Peugeot 205 GTi	1569
23	Audi 100	2119
24	VW Jetta	1577
25	Audi 100	1897
26	Toyota Tercel	1442
27	Opel Rekord 2,2i Autom.	2182
28	Fiat Regata	1585
29	Opel Kadett	1300
30	Toyota F (Category N1)	1984
31	Ford Fiesta Autom.	1100
32	Ford Fiesta M5	1100
33	DB 190 D	1983
34	DB 300 D	2996
35	Opel Senator 2, 3D	2300
36	DB 250 D	2479
37	VW Golf D	1570

* efficiency of the emission control system too low
** pilot run vehicle

COSTS

As will be apparent from the above discussion, the costs of meeting the proposed future regulations (Luxembourg Agreement) will depend very much on the size bracket of the engines in the vehicles. The three ranges will be considered separately.

For the additional costs a new baseline is selected: i.e. that of vehicles equipped to meet R15.04 regulations. The reason for this is to address the main objective of the report namely–what can be done to reduce acid pollution? And what will be the cost? As most vehicles are now capable of meeting R15.04, this seems to be the appropriate datum.

Typical additional costs (ex works) are given below:

Small Engine Vehicles (less than 1.4 litre)

1st Alternative: It may be possible with some vehicles to meet the regulations with engines tuned for R15.04 with the addition of EGR (Exhaust Gas Recirculation) and a moderately 'lean-burn' fuel setting (A/F ratio about 18/1). Some ignition retardation may be necessary. The latter would offset the improvement in fuel consumption obtained with moderate 'lean-burn'.

Additional Cost (EGR system)   c. £30

2nd Alternative: It is the intention of most manufacturers to reduce nitric oxide by a full 'lean-burn' system without EGR. This will require the following equipment:

(1) Programmed ignition with Lambda (λ) Sensor.

(2) Programmed Single Point (i.e. Throttle-body) injection system including microprocessor and ECU.

Additional Cost   c. £190

The second alternative is considered to give between 5–8% improvement in fuel consumption over the 1st alternative.

Medium Sized Engine Vehicles (1.4–2.0 litre)

Consensus of opinion says that the more severe regulations (compared with the small engine range) will necessitate the addition of an oxidising catalyst to the 2nd alternative above.

Additional Cost of Equipment   c. £290

Large Engine Vehicles (2.0 litre)

For large engined vehicles there are three alternatives.

1st Alternative: A three-way catalyst and a full engine management system with TBI   c. £390

2nd Alternative: For the larger engines in this range and for high performance vehicles, as 1st alternative but using a multipoint injection system   c. £465

3rd Alternative: The largest engined vehicles will probably require two catalysts in series   c. £565

One manufacturer states that he expects to spend £60 million in development to meet the Luxembourg Agreement figures on the three petrol and diesel engine ranges. Additionally they consider £70 million capital investment will be required to produce the systems.

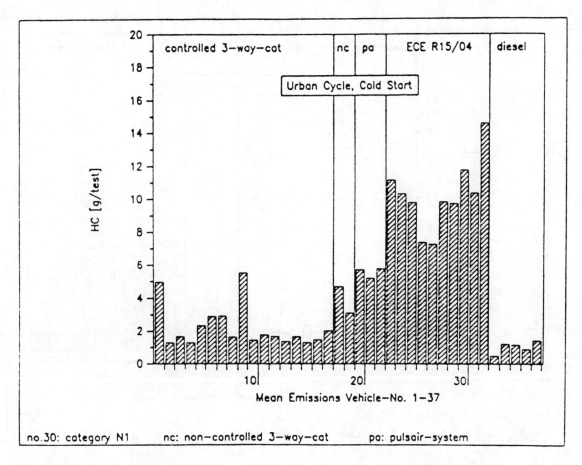

Figure 2.56 Comparison of HC Emissions in the ECE R15-04 Test from 37 Light Duty Vehicles Grouped According to Technology Class (TUEV Koeln & Essen)

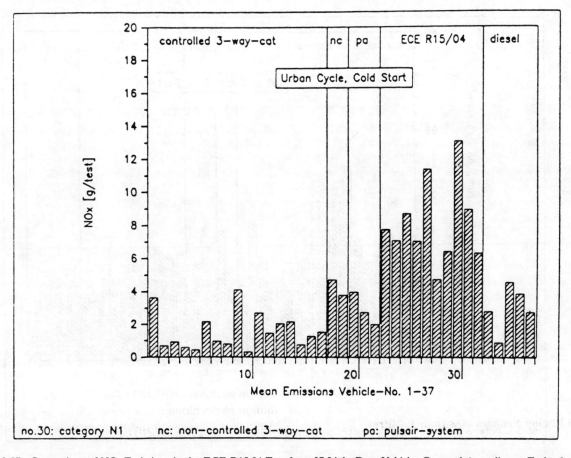

Figure 2.57 Comparison of NO_x Emissions in the ECE R15-04 Test from 37 Light Duty Vehicles Grouped According to Technology Class (TUEV Koeln & Essen)

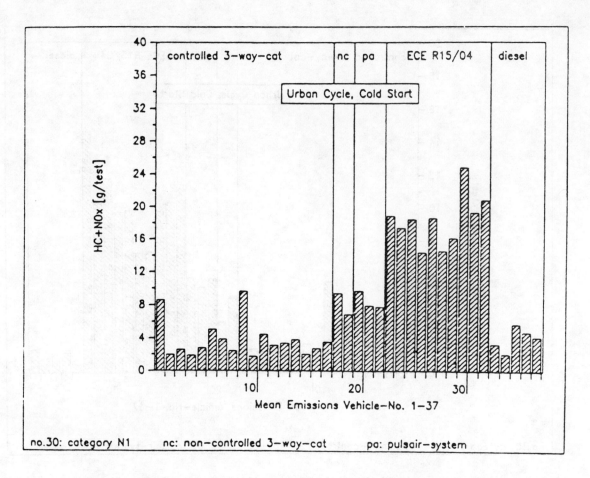

Figure 2.58 Comparison of HC+NO$_x$ Emissions in the ECE R15-04 Test from 37 Light Duty Vehicles Grouped According to Technology Class (TUEV Koeln & Essen)

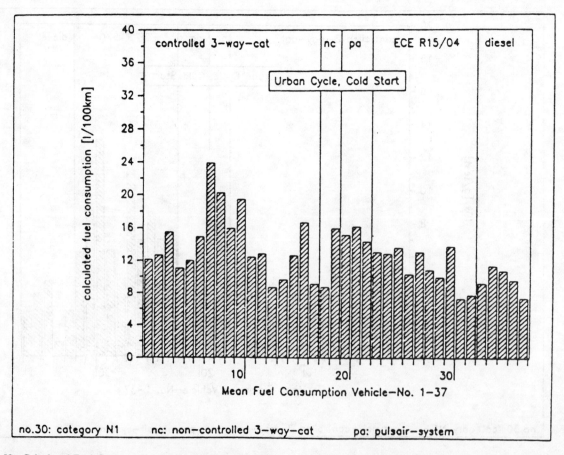

Figure 2.59 Calculated Fuel Consumptions in the ECE R15 Test Cycle for 37 Vehicles Grouped According to Technology Class (TUEV Koeln & Essen)

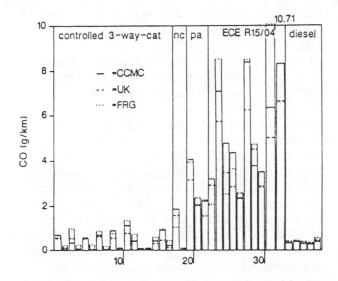

Figure 2.60 Comparison of CO Emissions in the High Speed Driving Cycles for 37 Vehicles Grouped According to Technology Class (TUEV Koeln & Essen)

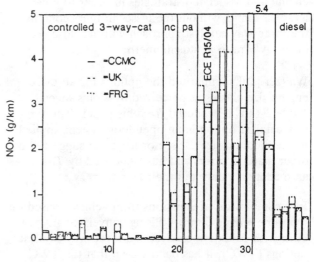

Figure 2.62 Comparison of NO_x Emissions in the High Speed Driving Cycles for 37 Vehicles Grouped According to Technology Class (TUEV Koeln & Essen)

maximum can be met by current technology similar to the strategies adopted to meet the ECE R15-04 regulations. For example, the major UK and European manufacturers' approaches are (i) attention to tolerances, (ii) improvements to fuel the ignition systems, and (iii) some management of ignition by the use of electronic control units. The fuelling systems used are mainly carburettors but some single point and multipoint fuel injection are also used.

The second stage of the regulation will be more stringent in the small car class and, for example, Ford expect to develop the lean-burn approach further with enhancements to engine control technology including adaptive calibration to optimise emissions, economy, power and driveability. It is generally agreed by manufacturers that the perceived Stage II will require oxidation catalysts.

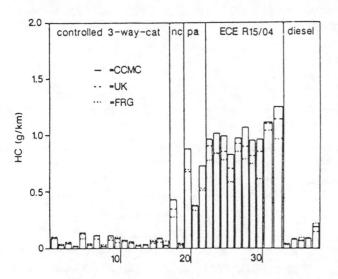

nc: non-controlled 3-way-cat   pa: pulsair-system

Figure 2.61 Comparison of HC Emissions in the High Speed Driving Cycles for 37 Vehicles Grouped According to Technology Class (TUEV Koeln & Essen)

SUMMARY

The proposed limits on emissions from new vehicles in the Luxembourg Agreement represent a considerable reduction in regulated pollutant emissions. For example, the contribution by large cars on nitrogen oxides reduction from the ECE R15-02 regulation is from about 15 g test^{-1} to 3.5 g test^{-1}; however, it should be noted that these figures are only approximate because the method of measurement and vehicle classifications have been altered.

In the small car sector (less than 1.4 litres) the proposed limits of 45 g CO, 15 g HC+NO_x with 6 g NO_x

The medium engine size sector (1.4–2.0 litres) has agreed limits of 30 g CO and 8 g HC+NO_x test^{-1}. Manufacturers are pursuing similar emission control approaches but tending to concentrate on the lean-burn strategy, which was the objective of the EC Member States in reaching the Luxembourg Agreement. Generally, lean-burn with an oxidation catalyst to meet the HC+NO_x limit is the preferred solution. The fuelling systems will be carburettors and single point fuel injection; some multipoint injection may be employed.

In the case of large cars equipped with engines greater than 2 litres European motor manufacturers agree that in the short to medium term the three way catalyst approach is the only practicable method of meeting the 25 g CO, 6.5 g HC+NO_x with 3.5 g NO_x test^{-1} requirements of the Luxembourg Agreement. However, it is clear that *Austin Rover*, *Ford* and *Volkswagen* are funding research into the lean burn engine and associated strategies in order to meet the large car limits without the losses in fuel economy of the three-way catalyst approach with its close control of the A/F ratio at stoichiometric.

Waltzer [10] summarised the options for all three engine size classes as perceived by Volkswagen. Table 2.14 is reproduced from his paper. The three-way catalyst option in an open loop system applied to the medium size car sector has been suggested by other manufacturers and has been used by *Toyota* as an oxidation catalyst in their lean burn system.

In a comparison of emissions from vehicles based on the same engine family utilising conventional, lean-burn with oxidation catalyst and three-way catalyst options (TWC), it has been shown that the TWC approach controls emissions to very low levels. The lean-burn engine plus oxidation catalyst will also control CO and HC to broadly similar levels; in the case of NO_x although the control achieved was 50–60% less emissions than its equivalent 'European' engine calibration it should be noted that the TWC control was in the range 88 to 95%.

The costs for increasing control of emissions vary but in general the current ECE R15-04 controls as compared with basic engine technology are £90–120, Luxembourg standards with full lean-burn plus oxidation catalyst would amount to £300–£400 and three-way catalyst systems plus full engine management would amount to £500–£600 all at 1987 prices. It should be stressed, however, that the price charged to the customer would be influenced by commercial and marketing considerations in addition to the actual costs incurred in production and development.

Table 2.14 Volkswagen Technology Options to Meet the Luxembourg Agreement

Displacement	1.4 l		1.4 l to 2.0 l		2.0 l
Euro Standard g/test					
CO	45		30		25
$HC+NO_x$	15		8		6.5
NO_x	6				3.5
Excess-air factor	1.1–1.2	1	1.4	1	1.6
Mixture formation	Two stage carburettor	Two stage carburettor	Single point injection	Single point injection	Sequential injection
Ignition	Management	Conventional	High performance	Map controlled	High performance
Exhaust treatment	Exhaust gas	3-Way cat.	Oxi. cat.	3-Way cat.	Oxi. cat.
-Sensor	–	–	Lean -Sensor	= 1 Sensor	Lean -Sensor
Swirl generation	–	–	–	–	Variable swirl
Fuel consumption improvement versus = 1	3%	0%	4–6%	0%	5–8%

3. Emissions Control Technologies – Diesel Engines

3.1 General Introduction

3.2 Fuel Injection Systems and Control Devices

3.3 Light Duty Engines

3.4 Heavy Duty Engines

3. Emissions Control Technologies – Diesel Engines

3.1 General Introduction

Diesel engines differ from Spark Ignition (SI) engines in that combustion is not initiated by a spark but by injecting the fuel directly into the cylinder in which air has been compressed to a degree which raises the temperature to a level sufficient to cause self-ignition. The injection is in the form of a fine spray, the droplets from which start to vaporise and mix with the entrained air. There is a short period, the delay time, before the ignition actually occurs and starts to increase the cylinder pressure further.

The diesel engine principle has many consequences. First, high compression ratios ranging between 15:1 and 22:1 are essential to give cylinder temperatures high enough to cause self-ignition. Secondly the fuel needs to be easily ignitable to keep the delay time as short as possible. This aspect of fuel is defined by the Cetane Number (see Section 4.2) which in practice ranges from about 50 down to 40, the higher the figure the shorter the delay time. Cetane Number is roughly the opposite to Octane Number as used for gasoline which increases with increasing resistance to self-ignition. Thirdly, the mode of combustion is such that complex cracking reactions can occur particularly in the larger fuel droplets in a spray and result in a greater risk of residual carbon particles which are not completely burned before being exhausted from the cylinder.

The high compression ratios have both advantages and disadvantages in relation to the SI engine. The ideal efficiency curve shown in Figure 3.1 shows that efficiency increases with increasing compression ratio. However because of the much higher internal pressures, the whole engine structure has to be stronger and heavier and the higher rate of pressure change increases combustion noise. Also the need to maintain good atomisation of the fuel spray over a wide range of fuel flows demands high injection pressures which in turn necessitate high precision and costly fuel injection pumps. These limit the engine speeds and cause an increase in engine size for a given power output.

However, the improvement in fuel economy of the diesel over the SI engines is increased by another factor. In the diesel engine, reduction in output is achieved by reducing the amount of fuel injected, whereas in the SI engine it is achieved by throttling

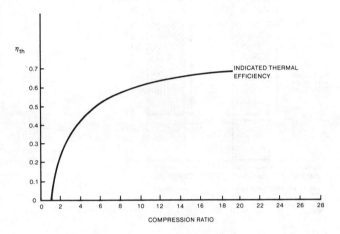

Figure 3.1 Variation of Engine Efficiency with Compression Ratio

the intake charge. This throttling at part load causes a significant amount of negative work which reduces the cycle efficiency. The energy balances for SI and diesel engines were discussed by Blackmore [11] who pointed out that at full load the difference between the two engine types is not great but at part load much greater differences appear, Figure 3.2 illustrates this.

One further factor in the comparison is that in the diesel engine a maximum of about 90% of the air can be utilised because of the difficulties in obtaining adequate mixing of the fuel and the air and this again adds to engine size for a given power output. However, turbocharging or some other form of supercharging can be applied relatively easily to the diesel engine to compensate for this, and the rapid increase in the application of turbochargers to diesel engines is a result.

Overall, therefore, the justification for the diesel engine so far is its better fuel economy against a number of disadvantages. It is thus understandable that its main application is in the commercial and passenger transport fields where the maximum economy is essential.

There are two major types of diesel combustion systems, the direct injection type (DI), Figure 3.3, and the indirect injection system (IDI), Figure 3.4. In DI engines, as the name implies, the fuel is injected

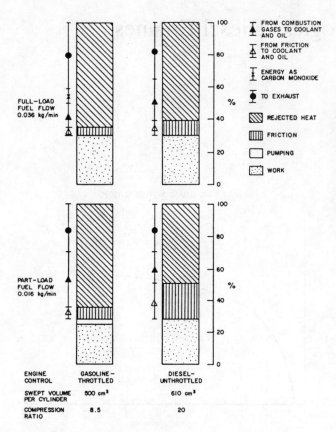

Figure 3.2 Energy Balance for Engines (2000 rev/min) (Blackmore and Thomas)

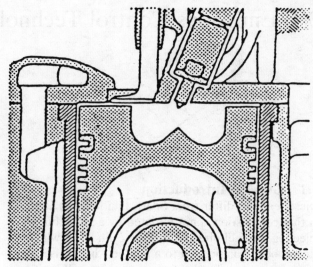

Figure 3.3 Swirling Direct Injection Combustion Chamber (Deep Bowl)

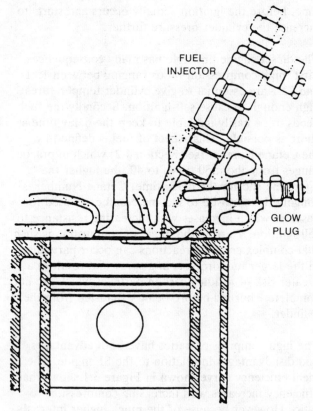

Figure 3.4 Cross Section Through Typical Indirect-Injection Diesel Engine

directly into an open chamber normally contained mainly in the piston crown. The IDI engine has a separate antechamber in the cylinder head into which air is forced as the piston rises and where the fuel is mixed with the air.

The antechamber of the IDI engine has an entry throat which produces a high level of swirl and turbulence which allows good mixing of the fuel and air without making severe demands on the injection system. This approach allows operation at higher speeds than in the case of the DI engine. It is also easier to apply in the case of small cylinders, where penetration of the spray can be difficult to control in DI engines. The penalties paid for these advantages are increased gas pumping losses in the throat of the antechamber and increased heat losses to the cylinder head. These result in some loss of efficiency.

Thus while the IDI engine has been easier to develop and apply particularly in the smaller high speed applications, the DI engine has the better fuel consumption. The latter is thus almost universally used in the larger applications and the incentive to develop it into the smaller high speed applications is great.

The level of this improvement in fuel consumption on the DI engine has been investigated in detail by *Ricardo* taking account of all the factors, injection timing and rate of heat release of the two systems as well as pumping and heat losses. Their results are summarised in Figure 3.5.

A large number of different configurations of combustion chamber and injector are in use in DI engines. An interesting development is the M combustion system developed initially by Dr J.S. Meurer of *MAN*. In this (Figure 3.6), the fuel spray is directed downstream into a swirling flow in a re-

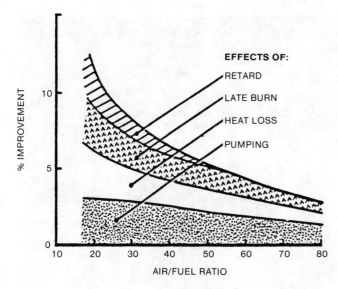

Figure 3.5 Improvement of Indicated Efficiency of DI Combustion System over Swirl Chamber

entrant combustion chamber in the piston and (over the upper part of the load range) impinges on the combustion chamber walls from which it evaporates and burns. This system gives good fuel economy and smoke limits and is noticeably quieter than most DI systems. It also makes it easier to match the injection requirements over a wide speed range. *Ricardo* have licensed this system and are developing it further, aiming particularly at the smaller DI engines. They term their development the Controlled Direct Injection System (CDI).

Another development aimed at increasing the speed range of the DI engine is the use of unit pump injectors (see Section 3.2). These eliminate the difficulties inherent in producing very high injection pressures using conventional distributor type injection pumps with delivery pipes to the injectors.

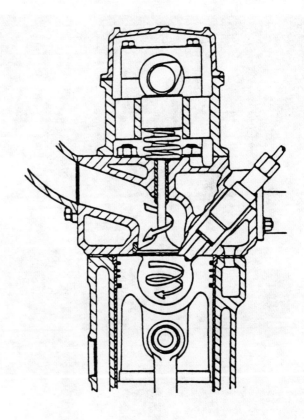

Figure 3.6 Meurer 'M' Combustion System for High Speed DI Diesels

As part of their work in this field *Ricardo* have made an extensive general assessment of three types of DI engine against the IDI engine for light duty vehicle applications with the results shown in Table 3.1.

Table 3.1 Functional Aspects of High Speed Direct Injection (HSDI) Systems (Compared with IDI of 4–6 1 cylinder)

	Conventional with pump-pipe-nozzle	Conventional with pump-injector	Wall-wetting–CDI system
Full speed power	5–10% < IDI	= IDI	= IDI
Low speed torque	5–10% < IDI	= IDI	5–10% < IDI
Urban fuel economy	10–12% > IDI	10–15% > IDI	5–20% > IDI
HC emissions	1–1.5 × IDI	1–1.5 × IDI	1–1.5 × IDI
NO_x emissions	1.5–2.0 × IDI	1–1.5 × IDI	1–1.5 × IDI
Partic. emissions	1–1.5 × IDI	1–1.5 × IDI	1–1.5 × IDI
Drive-by noise	2 dBA > IDI	2–2 dBA IDI	= IDI
Idle noise	greater than IDI	greater than IDI	less than IDI
Starting	10–20°C < IDI	10–20°C < IDI	10–20°C < IDI
Heat rejection	40–50% < IDI	40–50% < IDI	30–40% > IDI
Boosting ease	= IDI	> IDI	greater than IDI
Fuel tolerance	= IDI	= IDI	= IDI

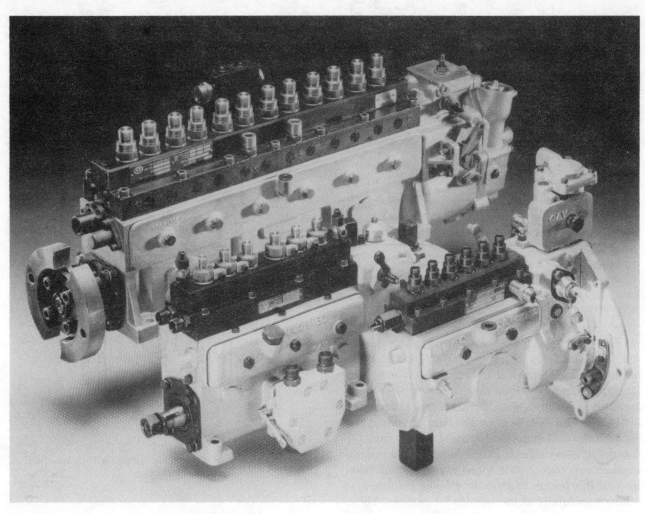

Figure 3.7 Lucas CAV In-Line Diesel Engine Pumps

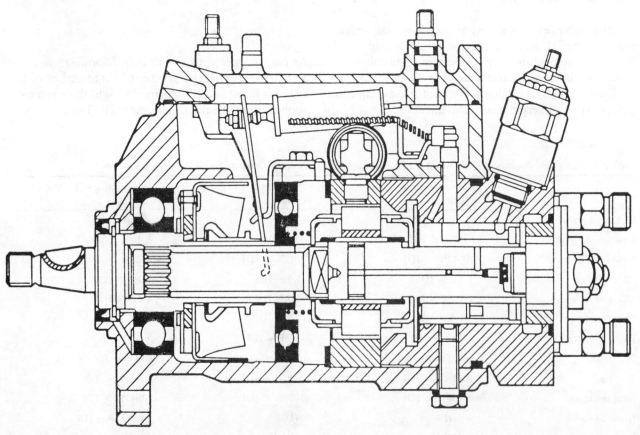

Figure 3.8 CAV Rotodiesel DPC Pump

3.2 Fuel Injection Systems and Control Devices

DIESEL FUEL INJECTION EQUIPMENT – GENERAL

With diesel engines, as with petrol engines, the fuel system plays an important part in relation to pollutant control. As will be seen in Section 3.3 nitric oxide emissions are very much a function of injection timing and spray quality. Because a considerable torque is required to drive the injection pump the varying of the injection timing is a much more difficult problem than the adjustment of spark timing in a SI engine.

A complete fuel system consists of a low pressure fuel pump to feed the injection pump, which in turn feeds the injectors with high pressure fuel accurately metered and timed. This high pressure fuel forces open the injector plunger giving a fine fuel spray.

There are three basic types of fuel pump:

(1) The In-line Pump: The pump has a separate piston for each injector each operated by its individual cam (Figure 3.7).

(2) The Distributor Pump: A *Lucas–CAV* DPC is shown in Figure 3.8 and consists of an internal cam operating a pair of pistons which come together forcing the fuel between them to a distributor which distributes the fuel to each injector in turn.

(3) The Unit Pump Injector: With this system injection pump and injector are combined (Figure 3.9), each plunger being operated by a separate cam (Figure 3.11). This eliminates the pipes connecting the pump to the injectors as in Types 1 and 2 and enables higher injection pressures to be obtained giving finer droplets of fuel as the volume and elasticity of the pipes are no longer present.

INJECTOR TIMING SYSTEMS

In Type (1) fuel pumps injection may be advanced by means of a simple but robust inertia weight mechanical governor which gives advance with speed; it is however unable to respond to load or electronic signals from sensors.

Type (2) has a greater degree of control as the internal cams can be moved relative to the drive thereby advancing or retarding the point of injection. In the CAV system the cam position can be controlled to provide the appropriate timing for starting, hot and cold idle and the various load and speed

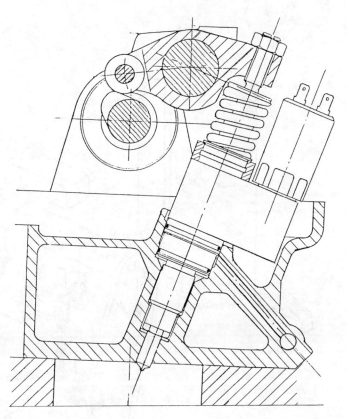

Figure 3.9 Lucas Unit Injector – Typical Installation

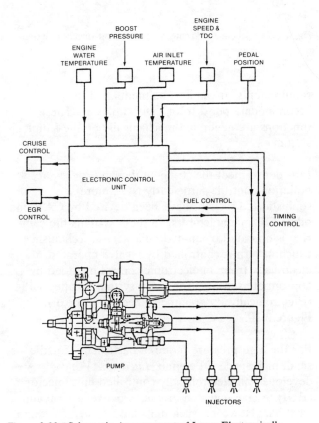

Figure 3.10 Schematic Arrangement of Lucas Electronically Programmed Injection Control System

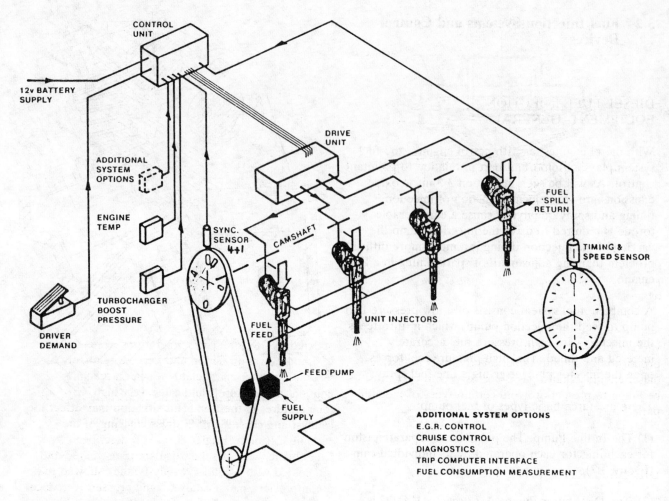

Figure 3.11 Schematic Arrangement of a Unit Injection System

requirements. It can also be modulated to accommodate boost from a turbo-blower. These functions are generated mechanically on the simpler system.

In order to meet the stringent demands necessary for pollution control, particularly NO_x, more sophisticated systems have been devised by CAV and other FIE manufacturers. In these systems the cam-ring is operated by an hydraulic servo-mechanism which in turn is controlled by a valve operated electrically from an electronic control box feed by appropriate sensors. The CAV system is called EPIC (Electronically Programmed Injection Control), Figure 3.10.

For the high pressure required particularly for the small high speed DI engine, *Lucas–CAV* have developed an electronically controlled unit injector (EUI) to give injection pressures between 1000 and 1200 bar. However, such units have to be an integral part of the engine. The total control system is shown in Figure 3.11. A high speed solenoid actuator (Colenoid) is used to control the valve which initiates and cuts off the fuel supplied by the cam operated plunger pump. The Colenoid is controlled by an electronic system, as above, in EPIC.

EXHAUST GAS RECIRCULATION SYSTEMS

Exhaust gas recirculation (EGR) is an important method of reducing nitric oxide in a diesel engine as a NO_x reduction catalyst is not possible in the prevailing oxidising atmosphere in the exhaust gases of a diesel engine. The effect of such recirculation is considered in the sections following. The method and control is similar to that described for petrol (Section 2.1) with one important difference. In the diesel engine the inlet air is not throttled and so there is only a comparatively low depression (caused by the air cleaner and some flow restrictions) to induce the exhaust gas (at a small positive pressure) into the inlet manifold. A bigger pintle valve therefore has to be used (Figure 3.12). Again, as there is no variable vacuum pressure signal available from the inlet manifold, the controlling pressure has to be generated from the engine driven vacuum pump usually available on trucks for brake system operations. To control this pressure for metering purposes a vacuum regulating valve has to be used. This is controlled by the position of the accelerator actuating the fuel system or by other more sophisticated means.

Owing to the presence of a small amount of acidic

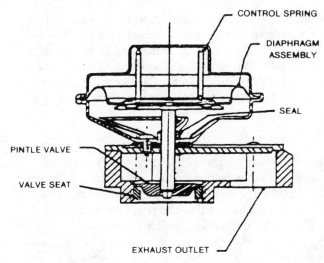

Figure 3.12 General Arrangement of a Diesel Exhaust Gas Recirculation Valve

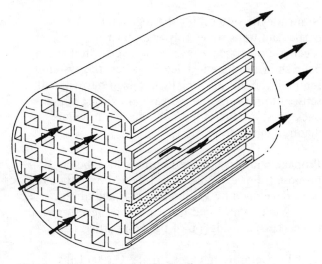

Figure 3.13 Diesel Particulate Trap–Ceramic Monolith

vapour, derived from the water, and a small proportion of sulphur dioxide and nitrogen dioxide, diesel engine exhaust gas is corrosive and care must be taken to protect or avoid certain types of metallic parts.

DIESEL PARTICULATE TRAPS

Though no completely satisfactory particulate trap (filter) has yet been produced, they have been proposed as a means of reducing to very low levels, which are well below present requirements in the USA (less than 0.12 g km^{-1}) emissions of particulate material from diesel engines. The principal trap designs are (i) ceramic monolith trap, (ii) ceramic foam, and (iii) alumina coated catalytic wire trap. Experimental traps have been applied to both light-duty and heavy-duty vehicles; however, the requirement to 'regenerate' the trap after the build-up of trapped particulate material was essential to the continued operation of the vehicle due to the increase in exhaust gas back pressure. Generally, the regeneration was effected by the combustion of the collected particulates. The combustion process may be non-catalytic or catalytic; in the latter case the unit is known as a Catalytic Trap Oxidiser (CTO).

The ceramic monolith filter principle is shown in Figure 3.13. The monolith was based on the ceramic structure used in catalyst units which are installed in the exhaust systems of gasoline engined catalyst cars. In the case of particulate traps ceramic plugs prevented the exhaust gas flow through the otherwise open structure and forced the gases through the porous cell walls. If the cell walls were coated with a suitable catalyst, then the ignition temperature of the collected particles was reduced and the trap reached an equilibrium stage between burn-off and collected particulate matter. This equilibrium was affected by low ambient temperatures which gave rise to a long warm-up period and hence a relatively high mass of collected particulates which on reaching ignition temperature could cause very high temperatures in the trap. This situation may affect the catalyst and cause thermal stress cracking of the ceramic monolith. Engler et al. [12] described experiments with diesel particulate traps coated and not coated with oxidation catalysts; they showed that a non-precious metal catalyst could be used to reduce the ignition temperature by 80°C (from about 483°C to 404°C), and, in combination with a precious metal catalyst on the gas outlet side of the trap simultaneous HC and CO oxidation could be achieved.

Rijkeboer [13] described experiments with a *Johnson Matthey* catalytic trap oxidiser, the design of which was based on a wire mesh coated with a washcoat and precious metal catalyst. A schematic diagram of the trap is shown in Figure 3.14. The catalyst formulation was designed to minimise the conversion of sulphur to sulphate. The vehicle used in the work was a city bus. It was found that the trap had a large thermal inertia but this property enabled the catalytic activity of the trap to continue to operate under engine idling conditions during stops.

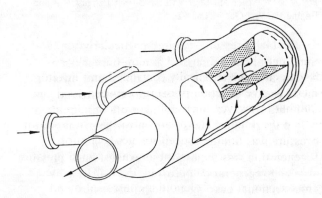

Figure 3.14 A JM Catalytic Steel Wool Filter

Non-catalytic systems for the regeneration of particulate traps which rely upon burner regeneration systems have been investigated and have been shown to be effective. However, the systems required the use of specially designed burners and sensors to measure the increase in exhaust system back pressure due to the build-up of particulate material in the trap.

Propane and diesel fuelled burners have been proposed. For example, diesel-fuelled burners have been developed by Wade et al. [14] and Simon et al. [15]. Wade et al. listed the conditions necessary for the successful oxidation of trapped particles as:

(i) Temperature in excess of 482°C (900°F).
(ii) Sufficient oxygen availability to support particle combustion.
(iii) Adequate time for completion of particle combustion.

Whereas sufficient oxygen would normally be available in diesel engine exhaust, and the combustion time can be engineered, diesel engine exhaust is usually at temperatures well below that required for ignition. An in-line burner was described [14] and is shown schematically in Figure 3.15. The maximum internal trap temperatures reached using this approach (with a mixing cone) was about 870°C. Burner light-off using spark ignition was a reliable procedure.

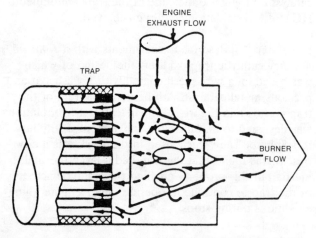

Figure 3.15 Schematic of an In-Line Burner Exhaust Gas and Engine Exhaust Gas Mixing Cone

Simon et al. [15] described a 'mini-burner' trap regeneration system in which a distributor swept a diesel fuelled flame across the inlet of a round monolith trap. Again, flame light-off was achieved with a spark plug. Measurement of the exhaust gas pressure was utilised to indicate when flame regeneration was required; the exhaust back pressure at which regeneration occurred was 6–7 kPa. A cross-sectional drawing of the 'mini-burner' and associated equipment is given in Figure 3.16.

Burner times varied according to the design (in-line or mini-burner) and stage of development. Thus for a 2.3 litre diesel engined vehicle and an in-line trap regeneration burner a 2 minutes regeneration time was required, whereas in the case of the 'mini-burner' with its flame distributor regeneration times of 5, 6 and 7 minutes were reported.

The catalytic trap oxidiser has the virtue of simplicity and if the temperature conditions can be optimised automatic regeneration of the trap occurs by the catalytic oxidation of the particulates. However, problems have included thermal stress damage due to high particulate loadings before automatic regeneration occurred. The ceramic monolith filter requires additional heat input before regeneration can occur; this regeneration cycle can be engineered to occur at regular and controlled particulate loadings as detected by an exhaust back-pressure sensor. Arai et al. [16] emphasised that high particulate loadings must be avoided in order to prevent thermal stress damage. Future developments are, therefore, likely to occur in the control of regeneration of traps by the application of burners or electrical heaters. In the case of burners the fuel penalty is about 1 per cent. If the catalytic trap oxidiser can be developed to automatically regenerate under all operating conditions at a specified maximum particulate loading then the CTO approach may become the preferred technology for the control of particulate emissions from diesel engines.

3.3 Light Duty Engines

GENERAL

As noted above, the small diesel vehicle engine (under 3 litres) has up till now been of the IDI type. This is mainly because the IDI engine has been capable of higher rotational speeds which are necessary since the installations have generally been designed in the first place to take gasoline engines. The objective of improved fuel consumption over the gasoline engine has been achieved particularly in the light duty applications.

The DI engine [17] however has the potential of improving the economy by a further 10–15% over the IDI engine and the development of DI engines with the necessary speed range is proceeding. These have been pioneered by *AVL* in Austria and *Perkins* in the UK and are being fitted by *Ford* and *Austin Rover* in light vans. With some modifications, these are soon likely to appear in passenger cars.

It is thus possible that the IDI engine will be replaced by the DI engine in all applications over the next decade, although this view is not universal. *VW*, in

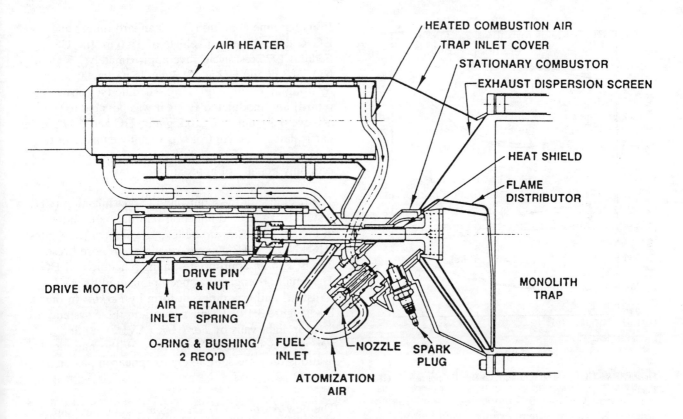

Figure 3.16 Cross-Sectional View of a Full-Sweep Diesel-Fueled Mini-Burner/Trap Assembly. Evaluated on both Vehicle and Dynamometer Installations of 6.2 litre Engines

particular, consider the small DI engine to be some 10 years away. Which type of high speed DI will become predominant is debatable but it is likely that the conventional type with the unit injector and the engines incorporating the CDI system will both be developed.

EMISSIONS AND FUEL CONSUMPTION

Nitric oxide and particulates are the predominant pollutants from diesel engines. Unfortunately catalytic reduction is not possible because there is always an excess of oxygen present. Additionally as the compression and hence expansion ratios are high, the temperature of the exhaust is considerably lower than in SI engines and this is not conducive to chemical reaction. As the production of nitric oxide in the combustion chamber is roughly proportional to temperature the only remedy is to reduce the temperature of combustion with the minimum loss of efficiency. There are two methods, exhaust gas recirculation (EGR) and injection retardation. However, other factors have to be watched, namely the effect of these palliatives on smoke and particulates.

The response to emissions control by EGR and injection retardation is well described by Ricardo [18]. Figure 3.17 shows the response of an IDI engine to both of these control measures—a light load operating point is chosen to illustrate their effectiveness over the statutory test cycles. It can be seen from the figure that the engine is tolerant to both EGR and retard and that the combination actually tends to improve all emissions with only small economy penalties.

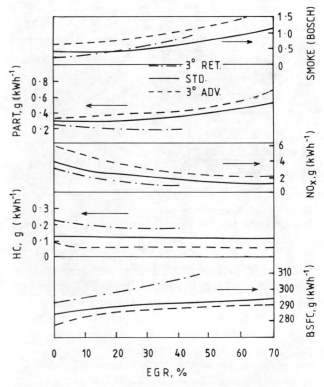

Figure 3.17 IDI at 2400 rpm and 3 Bar-Response to EGR and Timing

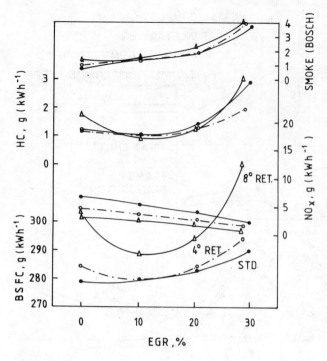

Figure 3.18 CDI at 3000 rpm and 3.5 Bar-Response to EGR and Timing

In its baseline condition, i.e. standard timing and no EGR, the engine in a vehicle of 3500 lb, the US Federal 1975 test cycle gave approximately 1.7 g/mile NO_x, 0.3 g/mile HC and 0.3 g/mile particulates, and by using an optimised timing schedule (selective retard) and modulated EGR it was possible to achieve 0.8 g/mile NO_x, 0.2 g/mile HC and 0.17 g/mile particulates. The control was roughly equivalent to 3 deg. retard and 15–20% EGR and resulted in an economy penalty of approximately 4%.

The response of the Controlled Direct Injection (CDI) system to EGR and injection retardation is shown in Figure 3.18. *Ricardo* believe it will meet the 8 g/test (HC+NO_x) and 2 g/test particulates proposed resolutions with an appropriate timing schedule and EGR. The small DI engine has been pioneered by *AVL* (Austria) and in a different form by *Perkins* in the UK (Figure 3.19). Engines based on these systems are fitted in light vans by *Ford UK* (AVL type) and *Perkins UK* in *Austin Rover* vans. With small modifications they are soon likely to appear in passenger cars.

The new *Austin Rover* DI engine recently announced (Type MDi–1.99 l capacity) has a novel combustion chamber design and was developed by *Perkins Engines* (termed the *Prima*). This is fitted currently in the *Austin Rover* Maestro Van. Test results show that the ECE R15-04 regulations are comfortably achieved but more work, which is in progress, will be required to meet the proposed 'Luxembourg Agreement' for HC+NO_x (see Table 3.2).

The problems with the small DI engine are noise and smoke. A high degree of spray atomisation is necessary to avoid wall wetting; this requires high pressures and attendant expense. The characteristics of the *AVL* system have been evaluated by Wade et al. [19] who made three experimental engines of different sizes.

Considering the emission characteristics, a quotation from their paper follows:

> 'To control HC emissions, compression ratios over 20:1 and fuel injection nozzles with valve covered orifices (VCO) were used. High compression ratios provided high end of compression temperatures which promoted combustion of the fringes of the fuel spray which were mixed to very lean local air/fuel ratios. The fuel injection nozzles with valve covered orifices prevented the after-dribble of unburned fuel into the chamber during the expansion stroke. In addition to improving HC emissions, these features also provided improved cold start and white smoke, reduced odour, reduced noise and lower peak firing pressures. Reduced NO_x emissions were obtained by retarding timing. Retarded timing without misfire was made possible with the features which provided low smoke and

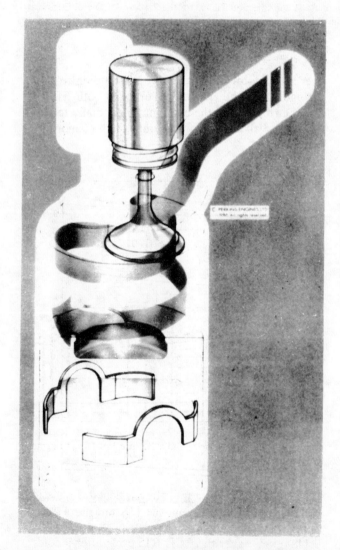

Figure 3.19 The Perkins Prima Combustion System showing the Helically Shaped Inlet Port for High Air Swirl

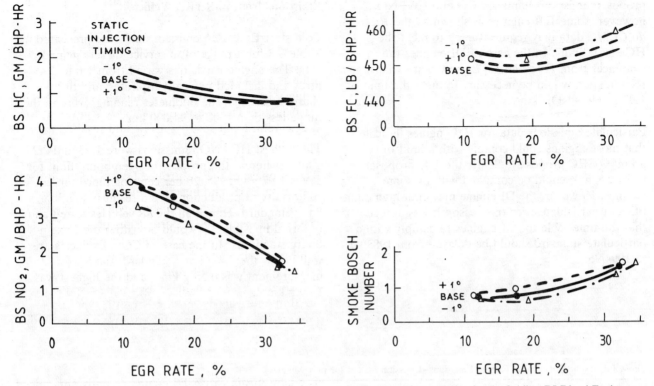

Figure 3.20 Emissions, Fuel Consumption and Smoke as a function of EGR and Injection Timing for the 2.4 litre DI Diesel Engine at the 1250 rpm/37 psi BMEP Condition

HC emissions. NO$_x$ emissions were also reduced with the high compression ratio which shortened the ignition delay. A shortened ignition delay reduced the amount of fuel which burned in the premixed mode where a significant portion of the NO$_x$ emissions is formed.'

To relate these emission characteristics to actual vehicle emissions the 2.4 litre naturally aspirated engine will be considered. The response to EGR and retardation is shown in Figure 3.20.

This engine was installed in a vehicle (3125 lb wt) and tested for emissions with a 2.3 litre IDI engine for comparison (Figure 3.21). The running cycle on the chassis dynamometer was the normal US-LA4 driving schedule. Exhaust analysis was by the Constant-volume-sampling technique.

Fuel economy was 11% better with the DI system. It should be borne in mind however that this was an experimental, not a production engine. Fuel pressures up to 1,000 bar were needed, which is beyond the capability of the distributor type pump employed if long life is required.

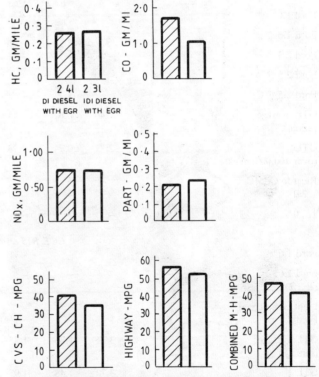

Figure 3.21 Vehicle Emission and Fuel Economy Test Results for the 2.4 litre DI Diesel Engine and a 2.3 litre IDI Diesel Engine

Emissions from New Vehicles

Type Approval (TA) data and TA estimations are given in Table 3.2 for ECE R15 tests. The Luxembourg proposed limits for gaseous emissions from light duty diesel cars are equivalent to 7.4 g CO and 1.97 g (HC+NO$_x$) km^{-1}. Apropos the particulates emissions, the European Commission, the UK and the FRG have proposed limits equivalent to 0.32, 0.27 and 0.148 g km^{-1} over the ECE R15 cycle after a cold start. Thus in the case of gaseous emissions Table 3.2 shows that IDI engined vehicles can readily meet the proposed emissions limits for CO and HC+NO$_x$; however, some IDI engines as shown by the *Ricardo/CCMC* data may require tuning to meet the HC+NO$_x$ limits. In the case of DI engines the evidences from *Ford* and *GTL* show that exhaust gas recirculation would be necessary to meet the limit of 1.97 g (HC+NO$_x$) km^{-1}.

Particulate emissions data for IDI engines indicate that most engines could comply with a limit of 0.32 g km^{-1} (CEC proposal); clearly the UK proposal (0.27 g km^{-1}) could be complied with by some engines. However, the DI engine may emit high quantities of particulates; for this reason it was agreed that the time-table for DI engines to comply with a particulate standard should be delayed from 1988/90 to 1994/96.

Emissions from In-Service Vehicles

Cold start ECE R15 emissions data are presented in Table 3.3 for a range of in-service diesel engined cars. The engine capacities varied from 1.6 to 2.4 litres and included naturally aspirated and turbocharged designs. The odometer readings were in the range less than 1000 to 94,000 km.

The highest HC+NO$_x$ emissions were 1.93 and 2.28 g km^{-1} whereas the proposed Luxembourg limit for TA is 1.97 g km^{-1}, a 20 per cent tolerance for conformity of production would result in a 2.36 g km^{-1} standard. Hence all the 28 vehicles tested and reported in Table 3.3 would be within the Luxembourg standards. In the case of CO all vehicles are well below the 7.4 g km^{-1} standard; the lowest measurement was 0.53 g km^{-1} and the highest was 1.59 g km^{-1}.

Table 3.2 Light-Duty Diesel Engine Emissions–Emission factors for new vehicles

Engine size, l	Emissions, g km^{-1}					Fuel Consumption l(100 km)$^{-1}$
	CO	HC	NO$_x$	(HC+NO$_x$)	Particulates	
ECE R15 tests: IDI Engines						
Ford 1.6	–	–	–	0.89	0.05	5.7
Ford 1.6	–	–	–	1.23	0.07	6.0
Ford 2.3	–	–	–	1.36	0.20	8.3
Ford 2.5 (NA)	–	–	–	1.58	0.27	8.9
Ford 2.5 (TC)	–	–	–	1.51	0.07	8.8
GTL average (initial)	2.46	0.62	1.11	1.73	0.3	–
GTL (over 100,000 miles)	2.71	0.74	1.23	1.97	0.3	–
Ricardo/CCMC						
2.0	–	–	–	0.74–2.2	0.12–0.59	
2.0	–	–	–	1.36–2.84	0.12–0.59	
ECE R15 tests: DI Engines						
Ford 1.6	–	–	–	2.02	0.20	5.1
Ford 1.6 (EGR)	–	–	–	0.94	0.15	5.2
Light commercial Ford 2.5	–	–	–	3.70	–	8.3
GTL average (initial)	3.21	1.36	1.85	3.21	0.62	–
GTL (over 100,000 miles)	3.70	1.48	1.97	3.45	0.62	–
GTL (EGR + injection retard)–initial	2.71	0.74	1.23	1.97	0.62	–
GTL (EGR + injection retard)–average over 100,000 miles	2.96	0.86	1.36	2.22	0.62	–
Perkins Prima* (NA)	2.5	1.6	2.1	3.7	0.70	–
Perkins Prima (TC)	10.7	1.26	2.28	3.54	0.38	–

EGR: Exhaust Gas Recirculation
GTL: Gaydon Technology Ltd
*   Also known as the MDi as fitted in the A.R. Maestro van

Table 3.3 Emissions from In-Service Light-Duty Diesel Engined Vehicles – ECE R15 Cold Start Tests

Engine Size, litres	Emissions, g km^{-1}					Fuel Consumption l(100 km)$^{-1}$
	CO	HC	NO$_x$	HC + NO$_x$	Particulates	
WSL Data						
1.6	0.90	0.18	0.70	0.88	–	6.53
1.6	1.06	0.15	0.76	0.90	0.086	6.68
1.6	0.81	0.36	0.87	1.23	0.14	–
1.6	1.28	0.49	0.73	1.22	0.25	–
1.8	0.96	0.28	0.70	0.98	0.18	–
2.3	1.41	0.52	0.85	1.93	–	9.68
2.3	1.60	0.43	0.91	1.34	0.30	9.72
2.3	0.87	0.20	1.45	1.65	0.19	–
Ricardo 86/1946						
1–3 (IDI)	1.20	0.24	1.06	1.30	0.31	–
Federal Republic of Germany Data 'RW' TUEV: 20 Car Survey						
1.6	0.87	0.19	0.77	0.96	0.16	–
1.6	0.93	0.42	0.77	1.19	0.12	–
1.6	1.04	0.33	0.74	1.07	0.14	–
1.6	1.52	0.54	0.70	1.24	0.19	–
1.7	0.68	0.11	0.79	0.90	0.11	–
1.7	1.16	0.16	0.91	1.07	0.16	–
1.8	0.94	0.31	0.72	1.03	0.13	–
2.0	0.65	0.09	0.79	0.88	0.20	–
2.0	0.61	0.12	0.89	1.01	0.19	–
2.0	0.80	0.18	0.83	1.01	0.25	–
2.0	0.85	0.18	0.91	1.09	0.28	–
2.3	0.75	0.12	2.16	2.28	0.15	–
2.4	1.59	0.28	1.01	1.29	0.09	–
2.4	1.27	0.21	0.98	1.19	0.34	–
Turbo-Charged Engines						
1.6	0.92	0.36	0.65	1.01	0.12	–
1.6	0.53	0.17	0.90	1.07	0.11	–
2.3	1.02	0.22	1.01	1.23	0.14	–
2.3	0.87	0.10	1.64	1.74	0.32	–
2.4	0.93	0.16	1.20	1.36	0.14	–
2.3 (comprex)	1.00	0.24	1.01	1.25	0.16	–

In the case of particulate emissions the range was 0.086 to 0.34 g km^{-1}. Only one car exceeded the CEC proposed standard (0.32 g km^{-1}) by 0.02 g km^{-1}, four cars exceeded the UK proposed standard (0.27 g km^{-1}) and fifteen cars exceeded the FRG proposed standard of 0.148 g km^{-1}.

Rheinisch-Westfalischer TUEV compared the average ECE R15 emissions of the same 20 cars with the emissions determined in the US 1975 Federal test procedure (FTP). The results are shown in Table 3.4.

The above results show that the higher average speed of the FTP-75 driving cycle (34.1 km h^{-1}) results in lower emissions than the ECE R15 cycle (18.7 km h^{-1}). The average CO and HC+NO$_x$ emissions per test in the case of ECE R15 were 3.8 g and 4.76 g respectively which are well below the Luxembourg standards of 30 g and 8 g respectively. Furthermore the average particulates emissions were 0.704 g test^{-1} which is above the FRG proposal of 0.6 g test^{-1} but below the UK 1.1 g test^{-1} proposal.

Four cars were studied by WSL over a range of on-the-road driving conditions using the mini-CVS sampling technique. The measurement of hydrocarbons under on-the-road conditions was not possible because an on-line flame ionisation detector (FID) would be required. A summary of the CO, NO$_x$ and particulate emissions is given in Table 3.5.

The emissions of carbon monoxide were very low throughout the range of driving conditions. The highest CO emissions occurred in urban drives but the range of mean values was only 0.53 to 1.06 g km^{-1} for average speeds of about 20 km h^{-1}. (Gasoline engined cars may be expected to emit 10 to 40 g km^{-1}

Table 3.4 Comparison of Light-Duty Diesel Engine Emissions in the FTP and ECE Cycles (average results)

Cycle	CO	HC	NO$_x$	Particulates
FTP g km^{-1}	0.61	0.16	0.79	0.154
ECE g km^{-1}	0.95	0.22	0.97	0.176
ECE g test^{-1}	3.8	0.88	3.88	0.704

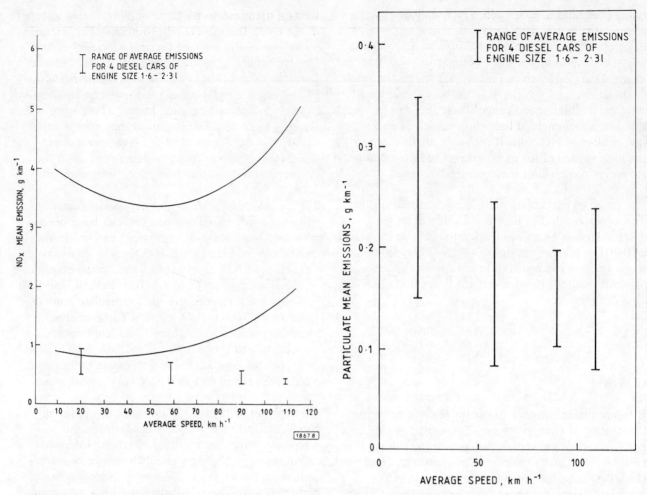

Figure 3.22 Variation of NO_x Emissions with Average Speed from Diesel Engined Cars compared with the WSL '20 Car Survey' Emissions Envelope

Figure 3.23 Variation of Particulate Emissions with Average Speed from Diesel Engined Cars on the Road

Table 3.5 Summary of Gaseous and Particulate On-The-Road, g km^{-1}

Test	CO	NO_x	Particulates	Average Speed km h^{-1}	Gear(s) used
1.6 litre vehicle					
Urban	0.70	0.50	0.35	20.0	All
Cross-Country	0.36	0.37	0.25	60.6	All
Motorway 90 km h^{-1}	0.26	0.42	0.20	92.4	4
Motorway 110 lm h^{-1}	0.32	0.35	0.24	111.1	5
1.6 litre vehicle					
Urban	1.06	0.74	0.16	18.3	All
Cross-Country	0.44	0.34	0.08	54.6	All
Motorway 90 km h^{-1}	0.42	0.45	0.10	86.3	4
Motorway 110 km h^{-1}	0.36	0.40	0.11	105.4	5
1.8 litre vehicle					
Urban	0.53	0.49	0.15	20.2	All
Cross-Country	0.26	0.29	0.08	58.5	All
Motorway 90 km h^{-1}	0.29	0.34	0.12	90.1	4
Motorway 110 km h^{-1}	0.18	0.38	0.08	108.2	5
2.3 litre vehicle					
Urban	0.65	0.94	0.22	22.6	All
Cross-Country	0.34	0.70	0.11	57.9	All
Motorway 90 km h^{-1}	0.30	0.55	0.11	89.9	4
Motorway 110 km h^{-1}	0.19	0.43	0.09	108.5	5

under comparable conditions.) NO_x emissions were also relatively low with a maximum average rate of 0.35 g km^{-1} in the case of a 1.6 litre vehicle under urban driving conditions; the comparison with gasoline engined cars is shown in Figure 3.22 where the range of diesel engined vehicle NO_x emissions are plotted over the average speed range studied. It can be seen that the diesel engined light-duty vehicle is also a low emitter of NO_x; this is probably attributable to the high air/fuel ratios used in diesel engines and hence the lower combustion temperatures achieved.

The range of particulate emissions measured on the road are presented in Figure 3.23. It can be seen that the highest emissions tended to occur in the urban drives. The highest emitter of the four cars studied was a 1.6 litre car which in urban drives emitted particulates at a rate of about 1.6 times that observed for a 2.3 litre car.

COSTS

It has not been possible to obtain realistic costs for the control of emissions from diesel engines (see also Section 3.4). However, in the case of light-duty vehicles EGR costs have been estimated by Ford (Europe) to be $40 for simple systems and $85 for modulated systems. These costs were stated to be valid for IDI and DI engine applications. In the case of particulate traps, such as was fitted to the *Mercedes 300D* for the Californian market, the cost may be £3–400; however, it should be noted that these traps have been withdrawn from that market.

SUMMARY

Light-duty diesel engines may be of the indirect injection (IDI) or direct injection (DI) designs. However, the majority of diesel engine designs in production for cars and vans are of the IDI type. In both types of engine the diesel fuel is injected during the compression stroke when, due to the compression of the combustion air, the temperature rises to a value which results in ignition of the fuel spray after a short delay.

In the case of IDI engines the fuel is injected into a small pre-chamber where a rich fuel:air mixture occurs and combustion is initiated. The hot gases then expand into the main clearance chamber; the volume of the pre-chamber may be up to 50 per cent or more of the total clearance volume. In order to control smoke and noise emissions the injection timing is usually retarded. The throat or duct leading to the pre-chamber causes some pumping losses and a loss of thermal efficiency.

In the DI engine, as its name implies, fuel is injected directly into the main combustion chamber. Hence losses due to the reduced thermal efficiency of the IDI engine and pumping losses are greatly reduced. Direct injection diesel engines such as the *Perkins PRIMA* may employ a bowl in the piston combustion chamber which reduces heat losses. This engine will also run at up to 4500 revolutions per minute rated speed. The development of the high speed direct injection engine was critically dependent upon the development of suitable fuel injection equipment.

The emissions of gaseous and particulate material from light duty diesel engined vehicles have been measured for a range of in-service vehicles. In the case of the cold start ECE R15 test the emissions of CO, HC and NO_x from naturally aspirated engines were in the ranges 0.61 to 1.6, 0.09 to 0.54, and 0.70 to 2.16 g km^{-1} respectively; the particulate emissions ranged from 0.09 to 0.34 g km^{-1}. Turbo-charged engines produced CO, HC and NO_x emissions, respectively, in the ranges 0.53 to 1.02, 0.10 to 0.36 and 0.65 to 1.64 g km^{-1}; the particulate emissions were in the range 0.11 to 0.32 g km^{-1}.

On-the-road tests over a wide range of operating conditions including urban, cross-country and motorway drives are reported. Carbon monoxide emissions were low throughout the range of driving conditions (about 0.5 to 1 g km^{-1}). Nitrogen oxides emissions were also relatively low with a maximum average rate of 0.35 g km^{-1} in the case of a 1.6 litre vehicle under urban driving conditions. Particulate emissions measured on the road were highest under urban driving conditions. The highest emissions of particulates measured from four cars were 0.35 g km^{-1} from a 1.6 l diesel engined car which was about 60 per cent higher than those measured from a 2.3 litre car under similar conditions.

It should be noted, however, that particulate emissions from direct injection engines may be considerably higher than those from IDI engines. For this reason the European Commission has agreed to a later implementation of particulate standards for these engines (1994/96) to allow for technical development.

Light duty diesel vehicles (cars, etc.) are covered by the 'Luxembourg' limits for the gaseous pollutants CO, HC and NO_x. Diesel engined vehicles of less than 1.4 litres are required to meet the same gaseous emissions standards as gasoline engines but *all* light-duty diesel cars of engine capacity greater than 1.4 litres are required to meet the gaseous standards for gasoline engines in the 1.4 to 2.0 litre range. Although not covered by the Luxembourg agreement, particulate emission standards for light duty diesels are being discussed. These are likely to be gravimetrically based and it seems probable that light duty vehicles will be able to meet the limits with current

IDI technology. Developments of light duty DI engines will improve fuel economy, probably by up to about 10%, but there will be a trade-off between decreased particulate and increased NO_x emissions.

3.4 Heavy Duty Engines

GENERAL

Practically all heavy duty diesel engines in commercial vehicles are of the direct injection open chamber type and this is likely to continue for some time.

Some small increase in rotational speed can be expected in the interest of improved power ratings but the major increase in power will come from turbocharging. This move has already started with an increasing proportion of the larger engines for heavy commercial vehicles being turbocharged. In most of these cases, an intercooler is incorporated between the compressor and engine induction manifold as this allows the power to be increased further and reduces the heat loading of the engine.

The effect of turbocharging on emissions is mixed. The increased oxygen available assists a reduction in smoke and particulates but the increased cycle temperature results in an increased amount of NO_x, although this latter effect is reduced by intercooling where this is used.

It is unlikely that there will be any further developments to affect this situation until the adiabatic or compound engine reaches the production stage which is probably a decade away.

EMISSIONS AND FUEL CONSUMPTION

The Economic Commission for Europe (ECE) Regulation 49 was adopted in 1982 at Geneva. However, the regulation has not been transposed into national or EEC homologation procedures. The CEC has put forward a proposal based on ECE R49 which requires reductions of 20% for CO and NO_x and 30% for HC from the ECR R49 levels. It is considered that the proposed new levels and somewhat different test procedures would bring European values to a level approaching the former US requirements for similar vehicles. This is a 13 mode steady state engine test (see Section 1.5). *Perkins* expect to meet these proposed regulations with a margin to allow for production variations by careful matching of the turbocharger to the engine requirements for their blown engines; modifications in combustion chamber design and some retardation of injection may be necessary. The latter could mean a 5% increase in specific fuel consumption.

To improve the trade-off between lower NO_x and fuel consumption *Perkins* have evidence that higher injection pressure would be beneficial; this is also supported by *Ricardo* and *AVL* (Austria). In order to achieve these high pressures (1500 bar) unit/pump injectors may be necessary.

The proposed new US regulations are even more severe and will inevitably require injection retardation. The test cycle, unlike the European, is a transient cycle for the vehicle, not the engine alone. It is necessary in the US to meet particulate standards and unfortunately reduction of NO_x tends to increase these particulates. *Cummins* have shown the trade-off between NO_x and particulates, Figure 3.24.

As it is likely that regulations for particulates will be introduced to Europe, it is clear that it would not be wise to trade increased particulates to reduce NO_x. It might be argued that particulate traps (see Section 3.2) are a possibility but no successful trap has yet been devised.

It seems clear from evidence examined that the emission levels required by ECE 49 can be met with only a small sacrifice to fuel consumption. To get substantially greater reductions fuel consumption will suffer and there will be an increase in particulate emissions.

Water injection, either into the eye of the impeller of the centrifugal compressor or into the inlet manifold is a very effective way of reducing NO_x but is not favoured because an additional tank to store water is required and there is risk of internal corrosion under certain conditions.

Only unproven expedients such as water injection, particulate traps, or adiabatic combustion systems are likely to give further improvement.

Emissions from Medium and Heavy Duty Vehicles

These vehicles may be broadly classified as vehicles of gross vehicle weights 3.5–16 tonnes and greater than 16 tonnes; buses and coaches are also considered in this category. Because emissions data are largely based on engine type approvals the data for whole vehicle emissions are limited.

Ricardo were commissioned by the CEC [20] to study the emissions of all diesel vehicles and make estimates of their contributions to the atmospheric burden of pollutants. In the medium goods class about 70% were naturally aspirated engines and 30% turbo-charged engines, whereas in the heavy goods area (over 16 tonnes) virtually all engines were turbo-charged. In 1983/4 about 35% of heavy duty truck engines were charge cooled but by 1986 the proportion was projected to be near 40%. The emissions estimated from the *Ricardo* study are given in Table

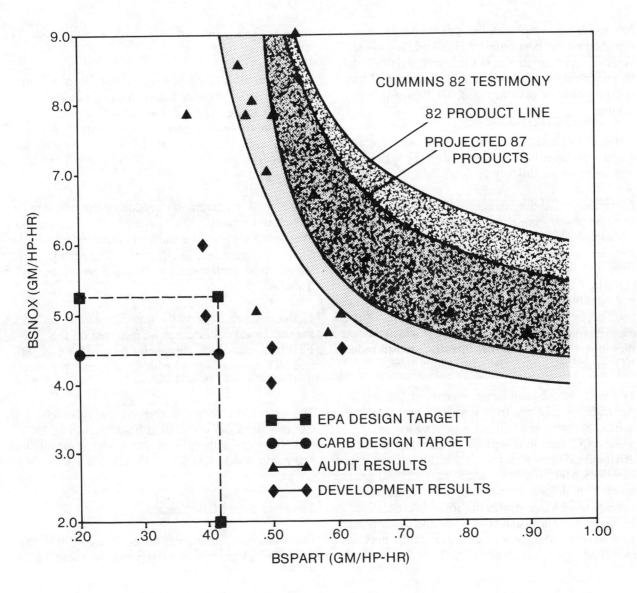

Figure 3.24 The Trade-Off Between NO_x and Particulates in the US Transient Test

Table 3.6 Emissions from Medium- and Heavy-Duty Diesel Engined Vehicles

Engine/Vehicle	Speed, km h^{-1}	Emissions, g km^{-1}					Fuel Consumption l(100 km)$^{-1}$
		CO	HC	NO_x	(HC+NO_x)	Particulates	
Ricardo report 86/1946							
Medium goods (NA)	65	3.41	0.61	6.58	7.19	0.55	)
Medium goods (GC)	65	2.00	0.57	13.07	13.65	0.37	) 16
Heavy goods (TC)	70	4.21	1.06	26.90	27.96	0.71	)
Heavy goods (TCA)	70	5.37	1.00	16.90	17.90	0.61	) 43
WSL road measurements (6 litre DI truck (NA))							
Unloaded	22.1	3.11	–	3.97	–	1.32	17.8
Loaded	22.8	4.10	–	3.70	–	1.70	20.1
Unloaded	63.0	1.53	–	2.10	–	1.21	13.9
Loaded	61.0	2.66	–	2.19	–	0.90	14.1
Unloaded	80	0.99	–	2.35	–	0.71	14.0
Loaded	80	1.97	–	3.08	–	0.65	15.9

NA: naturally aspirated
TC: turbocharged
TCA: turbocharged and after cooled

3.6. It must be emphasised that these data represented the best estimates available for in-use vehicles; verification was not possible practically but by comparing vehicle populations, activity and fuel consumptions an accuracy of about 10% was claimed.

Whereas CO and HC emissions for these large diesel engines are low, NO_x emissions are very high as compared with light-duty vehicles.

As mentioned above, turbocharging is used to increase power output and can reduce particulate emissions. Thus the *Ricardo* data for medium goods vehicles show a drop in average particulate emissions from 0.55 to 0.37 g km^{-1} for turbocharged as compared with naturally aspirated engines. However, it should be noted that the NO_x emissions are virtually doubled by the application of turbochargers. In the case of large engines charge cooling is widely adopted; it is shown in Table 3.6 that this strategy will reduce NO_x emissions substantially (by about 36%).

The WSL on-the-road measurements of CO and particulates emissions from a 6 litre diesel truck were carried out with and without a 1000 kg load. The highest CO emissions were obtained under urban driving conditions with the 1000 kg load; the range of emissions over an average speed range 22–80 km h^{-1} was 1.97 to 4.1 g km^{-1} which may be regarded as roughly in agreement with the *Ricardo* estimation. In the case of particulate emissions the experimental data were at variance with the estimations; thus the measured range was about 0.7 to 1.7 g km^{-1}, whereas the emissions estimated by *Ricardo* at an average 65 km h^{-1} were 0.55 g km^{-1}.

Ullman et al. [21] measured emissions from a 7 litre 2 stroke diesel engined bus on a chassis dynamometer using chassis dynamometer versions of the heavy-duty transient engine test procedure. An experimental bus cycle was also used in the emissions test programme. The two speed/time traces are reproduced in Figure 3.25.

The chassis dynamometer version of the transient engine test schedule was also used by Dietzmann and Warner-Selph [22] in a study of emissions from six heavy duty vehicles. The results of these studies are summarised, with respect to baseline emissions, in Table 3.7.

High emissions of carbon monoxide were observed for the 2-stroke 7 litre bus when tested on the FTP/EPA chassis dynamometer cycle. However, HC and NO_x emissions were within the ranges reported in Table 3.6 for UK estimations.

The six vehicle study also showed that relatively high CO emissions can occur from buses; however, HC and NO_x were within the ranges reported for medium and heavy goods vehicles in the UK studies.

Emissions from Static Engines

Emissions data for engines tested in accord with the ECE R49 13 mode test are reported in Table 3.8 for

Table 3.7 Emissions from Heavy Duty Engined Vehicles: US Procedures

Vehicle	Test	Emissions, g km^{-1}					Fuel Consumption l(100 km)$^{-1}$
		CO	HC	NO_x	(HC+NO_x)	Particulates	
Ullman Study							
7 litre bus (about 180 BHP 2 stroke engine)	FTP composite	53.6	1.56	10.2	11.76	4.4	47.6
7 litre bus (about 180 BHP 2 stroke engine)	Bus cycle	70.6	2.25	12.9	15.15	6.2	59.3
Dietzmann and Warner-Selph Study							
2-1 City bus	EPA chassis	21.4	1.74	10.8	12.54	1.28	–
2-2 Dual axle tractor	EPA chassis	5.56	2.06	14.3	16.36	0.97	–
2-3 Dual axle tractor	EPA chassis	2.24	1.72	13.4	15.12	0.87	–
2-4 Single axle tractor	EPA chassis	2.82	1.15	8.91	10.06	0.78	–
3-23 Single axle tractor	EPA chassis	3.70	3.16	8.99	12.15	1.19	–
3.24 Dual axle tractor	EPA chassis	4.67	1.62	17.6	19.22	1.35	–

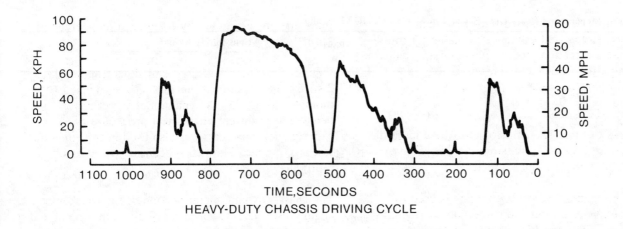

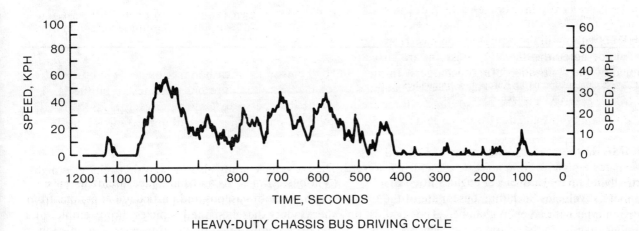

Figure 3.25 Heavy-Duty Chassis and Bus Driving Cycles

a range of *Perkins* Q20 Phaser engines and some FRG data supplied by the German Environment Ministry.

The 4 to 6 litre *Perkins* engines produced HC emissions rates that were below the CEC proposal of 2.4 g kWh^{-1}. However, the nitrogen oxides emissions were close to the proposed limit of 14.4 g kWh^{-1} and, in the case of the turbocharged T 4.40 engine, exceeded the limit (it should be noted that the *current* ECE R49 NO$_x$ standard is 18 g kWh^{-1}). The effect of charge cooling on nitrogen oxides emissions is demonstrated by the T 4.40 and T 6.60 engines; both engines when charge cooled produced about 19% less NO$_x$ than the uncooled turbocharged versions. Furthermore, the charge cooled engines produced more power due to the greater charge density achieved.

The range of data from the FRG for medium and heavy duty diesel engines is broadly in agreement with the *Perkins* data for engine powers in the range 60 to 130 kW. Furthermore it can be seen that large engines may emit levels of CO, HC and NO$_x$ that exceed the EC proposed limits. However, since 1986 there has been a voluntary agreement in Germany between government and industry that all new engines in series production will emit levels that are 20% below the ECE R49 limits, i.e. 11.2 g CO, 2.8 g HC and 14.4 g NO$_x$ per kWh.

COSTS

It has not been possible to produce any sensible figures for costs associated with various emission levels on diesel engines because these are linked closely with the basic engine design, in which there are many factors affecting the costs.

Even in the case of the main possible add-on component, the particulate trap, the specification and its development have not yet reached the stage at which realistic estimates can be made.

However, in the case of light-duty vehicles EGR costs have been estimated by Ford (Europe) to be $40 for simple systems and $85 for modulated systems. These incremental costs were stated to be valid for IDI and DI engine applications.

It must be remembered that the value of fuel consumed in the life of a commercial vehicle or a high mileage car is such that normal variations in engine cost are much less important than differences in fuel consumption.

Table 3.8 Medium and Heavy Duty Engine Emissions: ECE R49 Procedure

Engine/vehicle	Capacity, litres	Power, kW	Emissions, g kWh^{-1}		
			CO	HC	NO$_x$
Perkins Engines–Q20 Phaser					
4.40	4.0	65	–	1.1	13.5
6.60	6.0	98	–	0.95	14.5
T 4.40	4.0	79	–	1.26	17.1
T 6.60	6.0	119	–	1.14	13.8
T 4.40 CC	4.0	84	–	0.65	13.8
T 6.60 CC	6.0	134	–	0.66	11.2
FRG Data*					
up to 7 tonnes	–	60–130	2.7–7.0	1.4–3.0	8.0–14.0
more than 7 tonnes	5.6–15.8	90–235	3.0–13.0	0.8–3.0	10.0–21.0
CEC proposal			11.2	2.4	14.4

T : Turbo-charged
T----CC : Turbo-charged and charge cooled
CEC : Commission of the European Communities
* : FRG data for reference year 1980

SUMMARY

Virtually all heavy-duty diesel engines fitted in commercial vehicles (including buses) are of the direct injection type with an open chamber. The combustion chamber may be in the piston crown which results in reduced thermal losses. Furthermore, most large engines are turbo-charged thereby increasing the available oxygen and reducing smoke and particulate emissions; intercooling may also be employed in many designs in order to increase the charge density and, by lowering the charge air temperature, to reduce the formation of nitrogen oxides.

Although several national authorities (including the UK) in Europe are signatories to the ECE R49 set of emissions limits and test procedures, the Regulation has not yet been transposed into national or European Commission homologation procedures. However, Austria and Switzerland are in the process of mandating ECE R49 at levels 20% below the Regulation, and in the FRG a *voluntary* agreement has been made between industry and government that all new engine types will meet the 20% reduced levels. The test procedure is carried out on an engine test bed over a series of steady-state speed and load conditions and the results calculated on a mass per unit power basis. To date only gaseous emissions are regulated but particulate emissions will be considered in the future. In the United States a transient cycle has been introduced with effect from 1985.

Various estimations and some measurements have been made of gaseous and particulate emissions from heavy-duty diesel engined vehicles. For example, in a report commissioned by the European Commission *Ricardo* reported particulate emissions rates for medium and heavy goods vehicles ranging from 0.37 to 0.71 g km^{-1}; gaseous emissions rates were in the ranges 3.4 to 5.4 g CO, 0.57 to 1.06 g HC and 6.6 to 27 g NO$_x$ per kilometre for road speeds of 65–70 km h^{-1}.

On-the-road measurements of emissions for a direct injection 6 litre truck showed that the highest CO emissions (2–4 g km^{-1}) were obtained under urban driving conditions with a 1000 kg load. The highest nitrogen oxides emissions (3.97 g km^{-1}) with this vehicle were also under urban driving conditions with no payload. The measured range of particulate emissions was 0.7 to 1.7 g km^{-1}.

Engine test bed data for the Perkins Q20 Phaser engines are reported for 4–6 litre versions. Whereas the hydrocarbons emissions rates were well below the EC proposal of 2.4 g kWh^{-1}, the nitrogen oxides emissions were close to the 14.4 g kWh^{-1} proposed limit and in the case of the turbo-charged T4.40 engine exceeded the proposed limit at 17.1 g kWh^{-1}.

4. Fuels and Fuel Quality

4.1 Gasoline

4.2 Diesel

4.3 Liquid Petroleum Gas

4.4 Summary

4. Fuel Quality

4.1 Gasoline

GENERAL

Gasoline is a mixture of several hundred hydrocarbons which distil between 30°C and 210°C. The main components of gasoline fall into the broad chemical classifications: paraffinic, olefinic and aromatic. Furthermore, one or more of a range of additives may be included to modify the combustion characteristics to suit modern engine designs and air/fuel mixture preparation systems.

It is important that the fuel burns smoothly in the combustion chamber of a spark ignition engine so that the maximum amount of useful energy is made available. The phenomenon of auto-ignition (knock) limits efficiency and power and hence must be suppressed by suitable means so that the engine runs smoothly and mechanical damage is avoided.

The measure of a fuel to burn satisfactorily without knock in the combustion chamber of a spark ignition engine is known as its Octane Number. Two octane numbers are usually specified: Research Octane Number (RON) and Motor Octane Number (MON). In the arbitrary scale of octane numbers iso-octane is assigned a value of 100 and n-heptane a value of zero. The octane number of the fuel is then determined by comparing blends of iso-octane and n-heptane in an experimental engine until the knock performance of the blend is determined to be the same as the test fuel. The experimental single cylinder engines used for this purpose are known as CFR engines (Cooperative Fuel Research Council–USA). In practice the knock performance of the test fuel is bracketed by two iso-octane and n-heptane blends to determine the octane number accurately. The RON rating is determined using relatively light engine operating conditions whereas MON relates to more severe engine conditions at higher speed and load. The difference between the RON and MON rating of a fuel is known as its sensitivity and is an indication of the distribution of octane numbers throughout the boiling point range.

PHYSICAL PROPERTIES OF GASOLINE

Fuel Volatility

The fuel volatility at the 'front end', that is the first fractions to volatilise, affects the cold start performance of an engine, especially in low ambient temperature conditions. It is necessary for the fuel to volatilise sufficiently and pass in the vapour state mixed with air in the right proportions to the combustion chamber. The use of an enrichment device such as a choke is necessary to ensure a rich mixture for starting. The Reid Vapour Pressure (RVP) or other specifications, such as the amount of fuel evaporated at a certain temperature and pressure, have been correlated with cold start performance. The influence of gasoline volatility on engine performance is shown in Figure 4.1.

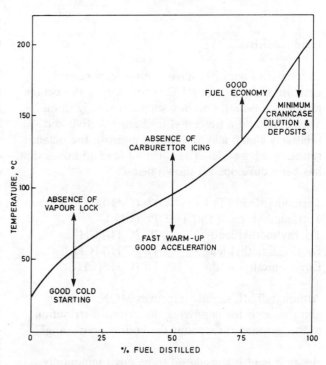

Figure 4.1 Influence of Gasoline Volatility on Car Performance

It can be seen that at the front end a careful balance between cold start performance and vapour lock is necessary. Similarly, the requirements for fast warm-up and carburettor icing avoidance, good fuel economy and minimum crank case oil dilution are important parameters. Thus changes in fuel composition which affect fuel volatility are carefully controlled at the refinery.

Density and Viscosity

The density and viscosity of gasoline are of significance to the air/fuel mixing process in, for example, the carburettor where the flow of liquid through orifices is affected. Thus the mass flow rate through an orifice (jet) is a function of parameters which include the square root of the fuel density and the coefficient of discharge of the jet which is a function of the Reynolds number:

$$Re = v\, Dd_f/\mu$$

where v = velocity of fluid flow
D = jet diameter
μ = viscosity of the fluid
d_f = density of the fluid

CHEMICAL PROPERTIES OF GASOLINE

As stated above, gasoline is a complex mixture of hydrocarbons; however, the final product will have certain additives which are designed to modify its combustion and chemical properties.

Lead Additives

Certain lead additives have been incorporated in gasoline since the early 1920's to improve the octane rating of the fuel, i.e. they suppress auto-ignition. Tetraethyl and tetramethyl lead are effective and relatively cheap additives which improve the octane rating of a base fuel. The range of lead additives that has been developed is shown below:

Tetraethyl lead (TEL)	: $Pb\,(C_2H_5)_4$
Tetramethyl lead (TML)	: $Pb\,(CH_3)_4$
Triethylmethyl lead	: $Pb\,(C_2H_5)_3CH_3$
Diethyldimethyl lead	: $Pb\,(C_2H_5)_2(CH_3)_2$
Ethyltrimethyl lead	: $Pb\,(C_2H_5)(CH_3)_3$

Although TML slightly improves MON, its main application is for improving the octane distribution of the gasoline throughout the boiling point range.

Because lead is considered to be environmentally unacceptable, agreements between the petroleum industry and governments have successively reduced the amount of lead in gasoline. Since the end of 1985 the United Kingdom has reduced the maximum allowable amount of lead in gasoline to 0.15 g litre^{-1}, representing a 60% reduction from the previous level of 0.4 g litre^{-1}. This reduction has been reflected in the atmospheric burden of lead; for example,

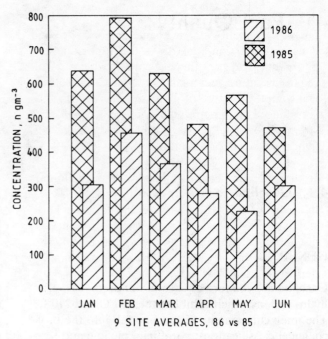

Figure 4.2 Plot of Monthly Average Lead Concentrations for 9 Sites for 1985 and 1986 Following the Reduction of Lead in Petrol to 0.15 g l^{-1} in January 1986

Williams [23] reported an approximate 50% reduction of lead in air in 1986 (see Figure 4.2). It should be noted that unleaded gasoline has a maximum lead content of 0.013 g litre^{-1}.

Other Additives

The decision by the UK Government to encourage the sale of unleaded fuel, and the commitment to make available unleaded fuel in accord with the EEC Directive 85/210/EEC by October 1989, has emphasised the requirement to maintain fuel octane quality. One approach is to increase the aromatic content (RON 100) of the fuel whilst observing the requirement of the Directive to maintain the benzene content to below 5% by volume. The changes of fuel composition and control of lead additives were discussed by Van Paassen [24]; he summed up the options for the refinery as:

1. More severe reforming
2. Use of more selective catalytic cracking catalysts
3. Upgrading of light naphtha
4. Alkylation and polymerisation
5. Use of oxygenates as blending components.

Johnson [25] discussed the Directive 85/536/EEC and concluded that future gasolines could contain up to 3% methanol, up to 10% other oxygenate mixtures, and MBTE (methyl tertiary butyl ether) up to 15%.

4.2 Diesel Fuel

Diesel fuel quality affects engine performance and emissions, and consequently in the motor and petroleum industries there is considerable discussion regarding future fuel quality standards. Diesel fuel is a complex mixture of many organic compounds and is produced from petroleum crude as a distillate straight run gas oil, catalytically cracked cycle oil, light residual oil or a blend.

In the UK, the petroleum industry has a voluntary agreement with the government to produce diesel fuel in accord with British Standard 2869 (1983). This standard calls for a 56°C minimum flash point, a minimum Cetane number of 50 and a maximum sulphur content of 0.3%. Other aspects such as density and viscosity are also specified.

The impact of fuel characteristics on diesel engine operation were summarised by the *Ford Motor Company* in a consultative paper (October 1985). Tables 4.1 and 4.2 are reproduced from that document.

The European motor industries Co-ordinating European Council (CEC) publish a specification for European legislative emissions testing. This is reproduced in Table 4.3.

Johnson [25] proposed a specification for diesel fuel which would be acceptable to the European motor industry as represented by the Committee of Common Market Automobile Contractors (CCMC). The specification, which is reproduced in Table 4.4, was considered adequate to 'preserve the efficiency and acceptability of their [manufacturers] engines'.

The Cetane number of a diesel fuel is a measure of its speed of ignition under the conditions of the compression stroke of the piston. Ignition delay occurs after injection due to the processes of fuel spray disintegration and droplet formation, evaporation of liquid fuel and diffusion of the into the surrounding air to form a combustible mixture. This ignition delay period is shown in 4.3 in terms of crank angle degrees. The correlation between Cetane number and ignition delay is shown in Figure 4.4. However, it is known that this correlation does not apply to non-petroleum based fuels.

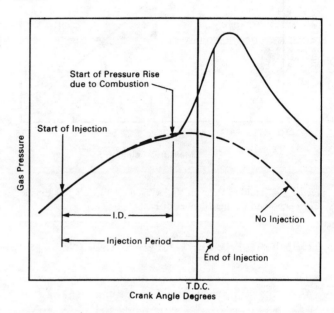

Figure 4.3 Cylinder Pressure in an Open Chamber Diesel Engine with and without Fuel Injection

Table 4.1 shows that the fuel Cetane number affects noise of combustion hydrocarbons and nitrogen oxides emissions, and fuel economy. Thus this aspect of diesel fuel quality is important to vehicle operation and the maintenance of emissions and acceptable noise in service. In a review of diesel engine emissions in Europe, *Ricardo* [26] plotted the distribution of diesel fuel Cetane number and Cetane

Table 4.1 Technical Influences on Diesel Fuel Quality Regulatory Requirements

Requirement	Influences	Fuel Parameter
Noise	– Ignition quality – Injection Timing	– Cetane number/index, volatility – Viscosity
Smoke	– Quantity of fuel injected – Timing	– Density, viscosity – Viscosity
Power	– Quantity of fuel injected – Timing	– Density, viscosity – Viscosity
Emissions		
– Gaseous HC	– Incomplete combustion – Nozzle coking	– Cetane number/index, volatility composition – Composition, stability suspected
NO_x	– High temp/high load	– Cetane number/index, viscosity
– Particulates	– As for HC, Smoke – Fuel constituents	– As for HC, Smoke – Sulphur, high boiling fractions
Fuel Economy	– Energy efficiency – Opportunities for Engine Technology Feasibility	– Density, calorific value, cetane number/index – Various

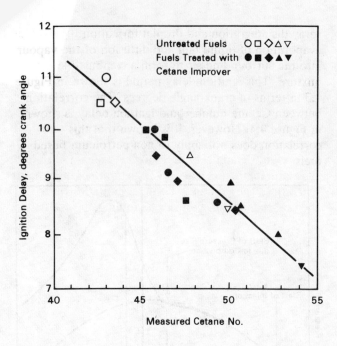

Figure 4.4 Ignition Delay in Volkswagen (VW) Golf IDI Engine Correlates with Cetane Number of Fuel

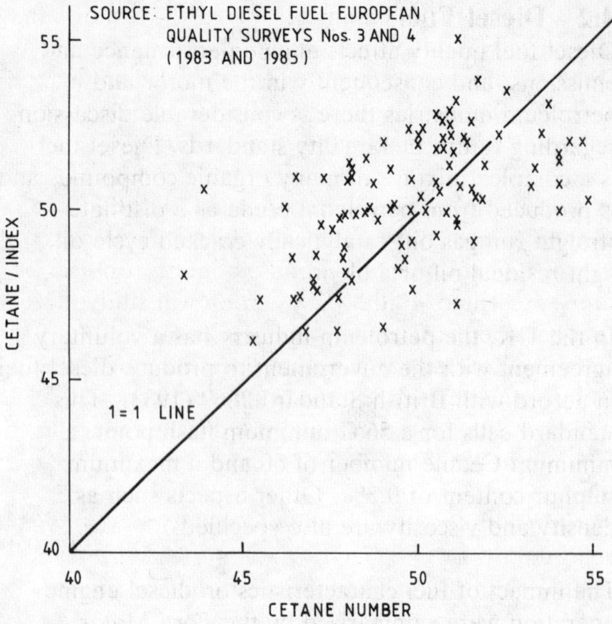

Figure 4.5 Results of Measured Cetane Number (ASTM D613) and Calculated Cetane Index (ASTM D976) for Samples of European Diesel Fuel

Table 4.2 Technical Influences on Diesel Fuel Quality–Consumer Satisfaction

Characteristic	Influences	Fuel Parameter
Cold operation		
– First start	– Ignition quality	– Cetane number/index, volatility
	Temp. related properties	– Volatility, CFPP, cloud point, viscosity
	– Nozzle coking	– Composition, stability suspected
– Cold start to clean run without smoke	– Ignition quality	– Cetane number/index volatility
	Nozzle coking	– Composition, stability suspected
– Sustained operation	– Filter clogging	– Waxing, viscosity, contaminants
Hot start/operation	– Internal pump leakage/no injection	– Low viscosity
Durability	– Pump wear	– Low lubricity, contaminants
	– Piston erosion	– Cetane number/index
	– Sticking nozzle needles	– Stability
	– Piston ring stick	– Ignition quality
	– Corrosion	– Water, lack of inhibitors/protection
Maintenance Intervals/costs	– Oil change frequency	– Hydrocarbon composition, sulphur content
	– Nozzle coking/cleaning	– Composition, stability suspected
	– Filter clogging	– Stability suspected, contaminants water content
	– Decarbonisation intervals	– Composition, stability, carbon content
Odour	– Exhaust smell	– Various
	– Fuel smell	– Various
Noise	– Ignition quality	– Cetane number/index, volatility
	– Nozzle coking	– Stability and composition suspected
Fuel economy	– Consistency	– Density variation
	– Absolute	– Density
Foaming	– Foam-back at filling	– Dispensing pump suitability Composition

CFPP = cold filter plugging point

index (a calculated parameter). Figure 4.5 shows that the range was 44–56. However, *Ricardo* comment that 'there seems to be no trend towards low quality fuel in any specific geographical area' and that the extremes noted were in the same country.

A working group has been convened by the Commission of the European Communities (CEC) in which governments, oil and automotive industries are represented. Initially, the group will study three fuel parameters: Cetane number, density and viscosity. One proposal is to reduce the minimum Cetane number to 48. (It is of interest to note that in winter in France the national standard is reduced from 50 to 48.)

A review of diesel fuel quality world-wide was reported in the October 1986 edition of *Petroleum Review*. Table 4.5 is reproduced from that review.

In the UK the Associated Octel Company [27] have published the results of their survey of diesel fuel quality which was conducted nationwide in 1986. Their results are reproduced in Table 4.6; from these data it can be seen that the average Cetane Number was 50.9 with a standard deviation of 1.2, the average density was 0.8456 with a standard deviation of 0.0047, and the average sulphur content by weight was 0.175 with a standard deviation of 0.06.

Sutton [28] investigated the effects of changing fuel properties with diesel operation and showed that gaseous and particulate emissions increased with decreasing Cetane Number. However, the addition of cetane improvers had only a slight effect on exhaust emissions. The addition of a detergent additive to the fuel reduced injector nozzle fouling.

4.3 Liquefied Petroleum Gas

GENERAL

Liquefied petroleum gas (LPG) has been used as a fuel for spark ignition engines for many years. In the United Kingdom the infra-structure for the supply of

Table 4.3 Specification for European Legislation Fuel for Emissions Testing

	CEC Reference Fuel RF-03-A-84 SPECIFICATION	
Type	Diesel	
Application	European Emission Tests (ECE 24, ECE 15) European Fuel Consumption Test (ECE 15) European Power Measurement Test (ECE 15)	
Property	Limits and Units (3)	ASTM Method (1)
Cetane Number (4)	min 49 max 53	D 613
Density 15°C (kg/l)	min 0.835 max 0.845	D 1298
Distillation (2) 50% 90% FBP	min 245°C min 320°C max 340°C max 370°C	D 86
Flash point	min 55°C	D 93
CFPP	min – max −5°C	EN 116 (CEC)
Viscosity 40°C	min 2.5 mm^2/s max 3.5 mm^2/s	D 445
Sulphur content	min (to be reported) max 0.3% mass	D 1266/D 2622 D2785
Copper corrosion	max 1	D 130
Conradson carbon residual (10% DR)	max 0.2% mass	D 189
Ash content	max 0.01% mass	D 482
Water content	max 0.05% mass	D 95/D 1744
Neutralisation (strong acid) Number	max 0.20 mg KOH/g	
Oxidation stability (6)	max 2.5 mg/100 ml	D 2274
Additives (5)		

For EEC Common Customs Tariff Regulations the 85% distillation point of this fuel must be specified on appropriate invoices.

Table 4.4 Proposal for a European Diesel Fuel Quality Definition

Characteristics	Test	Value Minimum	Value Maximum
Density at 20°C (kg/l)	D 1298	0.820	0.850
Viscosity at 40°C (cSt)	D 445	2.0	4.0
Cetane number, measured and cetane index	D 613 D 976/80	50.0 47.0	
Distillation	D 86	10% at 200°C 85% at 350°C Final boiling point	65% at 250°C 370°C
Sulphur % (Weight)	E.N. 7		0.15
Cold filter plugging point °C	I.P. 309 or CEN pr EN/16	Adequate to ensure operability in severe winter conditions.	
CFPP cloud point	I.P. 2500		10°C
Water content %, at point of sale	D 95		0.02
Sediments, at point of sale (mg/100 ml)			0
Flash point (°C)	D 93	55	
Ash %	D 482		0.01
Acid number (mg KOH/g)	I.P. 139		0.10
Storage stability (mg/100 ml)	D 2274		1.5
Copper Corrosion	D 974		1.0
Conradson carbon residue %	D 189		0.1

NOTE: A number of aspects of diesel fuel quality which are not included in this table are of concern for future Diesel fuel quality:
1. Aromatic content: it has a strong influence on particulates. The type and maximum content are to be determined.
2. Injector coking resistance: a test method and limit values should be determined.
3. Fuel foaming: disagreeable aspect at point of sale. Test and limit values to be determined.
4. Staining: vehicle bodywork stained by overflow. Test method and limit to be determined.
5. Calorific content: the customer is buying energy content. Limit values are to be established.
6. Seal integrity: future fuels may be more aggressive to fuel system seals. Tests and limit values to be established.
7. Biological contamination: preventative measures need to be taken to avoid fuel system clogging.

Table 4.5 Physical Inspection Data–Mean Values

Country	UK	West Germany	France	Italy	Benelux	Denmark	Spain	Japan	Singapore	Australia	USA	Canada
Number of Samples	8	18	16	12	7	4	2	13	6	6	12	6
Density (at 15°C)	.8460	.8332	.8378	.8411	.8367	.8376	.8472	.8406	.8473	.8540	.8532	.8336
Visc (cSt at 20°C)	5.09	3.60	4.25	4.69	3.95	4.11	4.83	5.54	6.95	4.95	3.99	2.94
Sulphur, wt %	0.18	0.17	0.25	0.48	0.17	0.33	0.41	0.44	0.29	0.25	0.18	0.14
Cetane Number	50.4	50.3	49.1	49.5	48.4	49.1	51.4	54.7	55.4	52.4	43.4	43.5
Cetane Index 1966	54.1	53.0	54.9	53.7	53.2	53.9	51.7	58.8	58.5	53.6	46.3	48.4
1980	51.1	50.6	52.1	51.0	50.8	51.4	49.3	54.6	53.7	50.8	45.0	46.3
Cloud Point (°C)	−1.8	−8.3	−1.0	+1.3	−7.7	−7.5	+0.5	−5	13.5	3.2	−14.3	−27.7
CFPP (°C)	−13.1	−19.3	−11.8	−12.0	−18.4	−17.5	−6.5	−10.2	+8.2	−1.0	−16.1	−30
Pour Point (°C)	−22	−23.5	−16.5	−24	−24	−23	−11	−15	+9	−3	−27	−34
Wax Content (Wt %) 10°C below cloud	2.6	1.6	1.8	1.8	2.0	1.8	2.1	4.4	3.8	7.7	2.5	1.0
D-86 Distillation (°C)												
IBP	180	167	165	172	158	173	176	188	190	192	179	168
5%	207	186	188	196	187	197	204	230	233	227	198	186
20%	240	213	217	225	217	224	234	257	266	249	227	206
50%	280	254	271	272	263	267	273	290	300	277	261	242
90%	339	327	347	349	327	330	337	332	364	330	317	296
FBP	362	359	377	378	354	362	363	356	387	356	349	328
90%–20%	99	114	130	124	110	106	103	75	98	81	90	90
FPB–90%	23	32	30	29	27	32	26	24	23	26	32	32

LPG does not exist. The number of refuelling points is about 1% of the UK's 25,000–26,000 service stations.

The supply of LPG (propane/butane mixtures) is increasing on a world-wide basis and has been predicted to increase from about 14.7 million tonnes in 1980 to 33.4 million tonnes in 1990 from Middle East and African sources and North Sea projects are predicted to reach 6–7 million tonnes in 1990.

The two European countries with most experience of LPG are the Netherlands and Italy. In both countries natural gas supplies had largely replaced LPG for industrial and domestic heating, hence the surplus of LPG created a reason for the promotion of the fuel for automotive purposes. In the Netherlands about 300,000 cars have been converted to run on LPG and about 15% of all mileage is fuelled by gas.

EMISSIONS

Fleming et al. [29] studied the emissions from a single cylinder CFR engine using gasoline and gaseous fuels (propane and natural gas). They also tested three US vehicles in accord with the 1972 CVS1 schedule in which after a cold start city driving is simulated.

In the case of the CFR engine experiments the most important conclusions are that the lean mis-fire limit is extended in the case of gaseous fuels and hence emissions of all three regulated pollutants (CO, THC and NO_x) may be minimised. The data are illustrated in Figure 4.6. It can be seen that in the case of CO the emissions are similar for the three fuels as the air/fuel equivalence (λ) is varied from rich to lean.

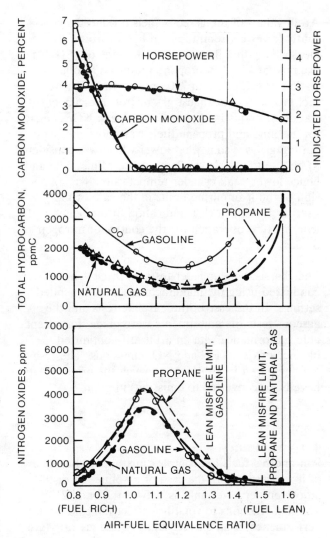

Figure 4.6 Power and Exhaust Emissions as a Function of Air/Fuel Equivalence Ratio at 50% Throttle for Gasoline, Propane, and Natural Gas Fuels. (Single-Cylinder Engine, Speed–1000 rpm; Ignition Timing 30° btc; Inlet-Mixture Temperature–160°F; 8 to 1 Compression Ratio; Intake Airflow–0.29 lb/min.)

Table 4.6 Associated Octel Survey of UK Diesel Fuel Property June 1986

Code Inspection Data	GB 1	GB 2	GB 5	GB 6	GB 7	GB 10	GB 11	GB 13	GB 14	GB 15	GB 16
Density, kg/litre @ 15°C	0.847	0.845	0.848	0.842	0.846	0.842	0.853	0.835	0.846	0.850	0.847
Viscosity, cSt @ 40°C	3.4	3.4	3.0	2.8	3.1	3.4	3.4	2.7	2.9	3.6	3.0
Cloud Point °C	+3	−2	+5	0	−2	+1	+3	+4	−1	+2	−1
CFPP °C	−11	−7	−3	−3	−13	−5	−4	−2	−5	−10	−10
Pour Point °C	−19	−13	−22	−13	−25	−13	−13	−13	−22	−25	−25
Flash Point °C	69	75	81	59	70	63	73	59	69	64	62
Distillation, IBP °C	177	181	182	164	183	172	180	171	183	168	173
5% vol °C	217	218	230	192	210	206	214	194	208	213	202
50% vol °C	291	288	294	278	284	290	284	271	278	290	281
65% vol °C	308	304	312	300	303	310	301	297	298	308	300
85% vol °C	340	331	342	324	335	345	333	341	330	339	337
95% vol °C	368	359	364	367	368	378	364	382	356	365	368
FBP °C	374	371	374	377	374	383	374	390	364	372	375
Sulphur, % wt	0.13	0.18	0.21	0.20	0.21	0.26	0.14	0.22	0.04	0.21	0.13
Calculated cetane index (CCI)	52.6	52.7	52.5	52.0	51.6	53.8	49.4	52.9	50.5	51.2	50.9
Cetane number	51.4	52.8	51.8	48.8	51.0	51.5	51.3	50.6	50.8	51.2	48.7

At all values of the gaseous fuels produced significantly lower concentrations of hydrocarbons as measured with a flame ionisation detector (FID). For example at $\lambda = 1.2$, propane produced a concentration of HC of about 750 ppm C whereas gasoline fuel produced an exhaust gas hydrocarbons concentration of about 1400 ppm C. The peak NO_x emission for gasoline and propane fuels was about the same; natural gas was somewhat lower which was considered to be due to the slower flame speed of methane and hence lower peak cylinder temperature. The extension of the lean misfire limit, in the case of the gaseous fuels, is a clear indication of the benefits of lean mixture operation for the control of nitrogen oxides.

The chassis dynamometer tests on three vehicles were conducted using the manufacturers' recommended settings. In the case of the gasoline tests and the gaseous tests the adjustments were made 'for acceptable performance with an arbitrarily optimised' balance of CO, THC and NO_x emissions. The results are compared in Table 4.7 in which the original data have been converted to emissions in g km^{-1}.

Table 4.7 Comparison of Emissions from Gasoline, Propane and Natural Gas Fuelled Cars

Fuel	Emissions, g km^{-1}		
	CO	HC	NO_x
Ford, 250 CID			
Gasoline	6.6	1.7	5.7
Propane	3.4	1.0	4.4
Natural gas	0.9	1.1	2.5
Chevrolet, 327 CID			
Gasoline	24.6	3.3	3.8
Propane	1.8	1.2	1.2
Natural gas	2.2	1.7	2.5
International, 345 CID			
Gasoline	11.2	2.7	6.6
Propane	1.7	1.7	1.4
Natural Gas	1.4	1.5	2.3

CID = cubic inches displacement

4.4 Summary

Automotive fuels are complex mixtures of hundreds of hydrocarbons. In the case of gasoline the distillation temperature range is between 30° and 210°C. The first fractions to volatilise affect the cold start performance of an engine. Density and viscosity are also important parameters of gasoline. Historically, lead compounds have been added to gasoline to improve its octane rating. The decision of the UK Government to encourage the use of unleaded fuel may lead to changes in fuel composition such as an increased use of oxygenates.

The boiling range of diesel fuel is from about 170°C to about 390°C. Fuel specifications and the effects of fuel properties on diesel operation and emissions are indicated. For example, fuel Cetane number affects combustion noise, emissions and fuel economy.

Alternative gaseous fuels are briefly discussed and the data from some early experiments show that propane and natural gas fuelled cars produced lower emissions of CO, HC and NO_x than the conventional gasoline version. Furthermore, CFR engine experiments showed that the air/fuel lean limit could be extended as compared with the conventionally fuelled engine.

5. Alternative Engines

5.1 Stratified Charge Engines

5.2 Miscellaneous

5.3 Production Possibilities

5. Alternative Engines

5.1 Stratified Charge Engine

It can be seen from Figure 2.1 that if an SI engine can be operated around a 22:1 air/fuel ratio then CO, HC and NO_x are all low; therefore, this is the approach adopted for the lean burn engine (see Section 2.3). The problem with the latter engine is to accelerate smoothly from this lean ratio to full torque and power. The stratified charge engine (SCE) is an alternative approach which does not suffer so severely from this problem, and because the SCE controls power by fuel supply, largely without throttling, it is more efficient. Three basically different approaches have been thoroughly evaluated. The pre-chamber type based on early work by the late Sir Harry Ricardo and produced, in modified form, by *Honda* and two open chamber types, the Texaco and the *Ford* PROCO. With the exception of the *Honda* no stratified charge engine has gone into production. This is probably due to the considerably greater complication and expense coupled with the fact that regulations have so far been met with more conventional engines. The SCEs still need an oxidising catalyst to reduce hydrocarbons.

OPEN CHAMBER STRATIFIED CHARGE ENGINES

As the direct injection (DI) diesel is superior to the IDI for fuel consumption it is considered that a direct injection SCE should be better than a pre-chamber type.

Ford PROCO Engine

The *Ford* (US) PROCO [30, 31] shown in Figure 5.1, is the most advanced example of this type of engine and differs from the Texaco in that it is not smoke limited. Injection of the fuel takes place near bottom dead centre (bdc) and the fuel evaporates by the time the sparking plug is fired. Injection is from the side of the chamber and the spray pattern has been developed to give a near stoichiometric ratio near the plug weakening to 20:1 A/F at the far part of the chamber.

Texaco Engine

The principle of the Texaco engine is illustrated in Figure 5.2. Air swirl is induced by the intake port in much the same way as in a diesel engine. Fuel is injected downstream into the air swirl where it evaporates and is ignited by a sparking plug. This forms a flame front or combustion zone into which the spray continues to be fed. Fuel injection is cut off and combustion continues until all the fuel is

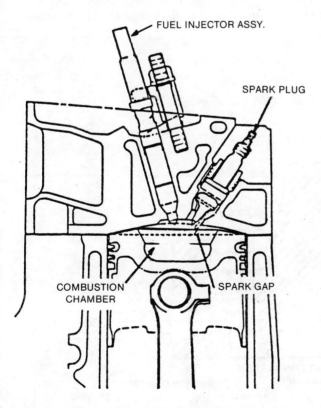

Figure 5.1 Ford 351-CID Proco Engine – Cross Section

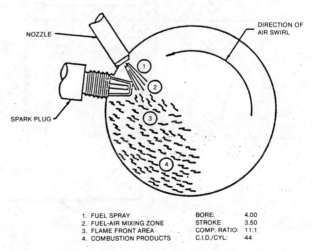

Figure 5.2 Texaco Engine Concept

consumed. As noted, development of this engine has proceeded sporadically over two decades but has not reached the production stage. The reasons for this will be considered in relation to its emission potential. It will burn a large range of fuels from methanol to diesel and like the diesel is smoke-limited, namely it is unable to consume all the air in the cylinder because of the onset of smoke. As it is unlikely to compete with the small DI diesel on either fuel economy or pollution, only the petrol version will be considered. It has to be said however that it is quieter than its diesel counterpart even when running on diesel fuel. With petrol because of the pattern of combustion, there is virtually no end gas, and knock propensity is much reduced enabling a high compression ratio to be used with a fuel of moderate octane number. This gives a considerable improvement in fuel consumption compared with a conventional petrol engine. Prototype engines have been manufactured and evaluated and two major projects may be quoted – the *Texaco* L-1635, a four cylinder engine manufactured by the *White Corporation*, and the UPS292SC engine manufactured by *Ricardo* for the 'United Parcels Service' [32] using the *Texaco* design and know-how in a 6 cylinder in-line version (4.8 litre), 13:1 compression ratio. This engine requires a high energy series of sparks for satisfactory ignition. Injection equipment is similar to that required with diesel engines.

Emissions

Emissions from the *Texaco* engine in a vehicle are shown in Table 5.1.

Table 5.1 Emissions from the Texaco-Engined Vehicle in the US 1975 Test

Vehicle inertia weight 2750 lbs:
Fuel – unleaded gasoline (Octane Number 91) with oxidising catalyst

	Emission g/mile			
	CO	HC	NO_x	Particulates
Test 1	1.05	2.75	1.2	0.1
Test 2	0.85	0.55	2.4	0.06

Test 2 was performed with a new catalyst

Although this engine was not specifically tuned to meet the latest US regulations is seems unlikely to be able to do so. The same is likely to apply to the proposed Luxembourg regulations figures for the over 2 litre cars and under this size the expense of the fuel and ignition equipment would not be justified.

PRE-CHAMBER STRATIFIED CHARGE ENGINES

This type of chamber, originally developed and researched by Sir Harry Ricardo to improve fuel consumption has been comprehensively investigated for its capability of reducing pollutants particularly nitric oxide. *British Leyland*, *GM* (US), *VW*, *Porsche* and Japanese companies have all built experimental versions of this type of engine [33]. *Honda* however, were the only company to go into production for their home and US market. At the time of this research work the additional expense was not justified for Europe. The regulations then (circa 1976) could be met with conventional engines, and the improvement

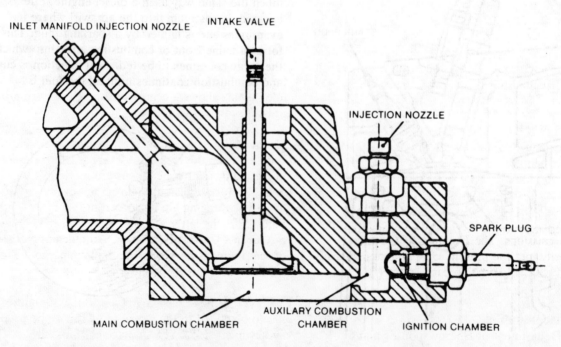

Figure 5.3 General Arrangement of the Porsche SKS Stratified Charge Engine

in fuel consumption though clearly demonstrated was not considered a sufficient incentive for the capital expenditure that would be required for a new engine.

Honda Engine

The basic design of the *Honda* engine consists of a small pre-chamber not dissimilar to that of an IDI diesel engine but is generally considerably smaller. This pre-chamber has its own mixture fed through an additional small inlet from virtually a separate carburettor. However, in the case of Honda this was combined in a single dual-carburettor design. A readily ignitable mixture, around stoichiometric, is fed to the pre-chamber where it is spark ignited in the usual manner and the burning gas forms a torch which rapidly discharges through the nozzle and ignites the main charge above the piston. The main charge from its separate air and fuel supply was capable of being adjusted in air/fuel ratio to meet the required needs.

Experiments demonstrated that this method of torch ignition was capable of igniting very lean mixtures well lean of 24/1 A/F ratio.

Porsche Engine

Porsche, Figure 5.3, performed a systematic variation in the size of the pre-chamber relative to the main chamber and demonstrated that as little as 1/10 ratio was about optimum. Orifice size is also significant as it influences the penetration of the burning jet and in consequence the burning pattern in the main chamber.

Emissions

A 4 cylinder *BL* experimental engine (1855 cc) in a Marina vehicle produced the following results for the US-CVC Regulations.

Emissions gm/mile	NO_x	CO	HC
US Californian Regulations 1976	2	9	0.9
BL Experimental SCE	0.7	6	8.0

It will be seen from these figures that NO_x reduction is good but HC poor. With better mixture control and an oxidation catalyst, such a vehicle should have very low emissions. The *Honda* CVCC has satisfactorily met the present US regulations because it is fitted with an oxidation catalyst.

5.2 Miscellaneous

The US Department of Energy sponsor many development contracts for alternative engines with the object of lower pollution and greater efficiency. They also organise symposia on a regular basis on automotive propulsion systems which give publicity to this work. The symposia are wide ranging covering many alternative engine systems as well as advances in the diesel engine and spark ignition engine. Additionally the Department sponsors research on several power systems through contracts, industrial firms and research organisations. Contractors meetings are held from time to time outlining the progress of work. Such meetings are open to all interested parties.

In this section which covers two-stroke engines, gas turbines, Stirling engines, Rankine engines, the US Department of Energy-sponsored work will have a prominent place because its objectives are very similar to those of this report.

TWO-STROKE ENGINES

General

The two stroke engine accomplishes the full cycle of events in two strokes or one engine revolution. This distinction from the common 4-stroke engine has pollution implications. Ports or valves open as in a 4-stroke engine some 70° before bottom dead centre (bbdc). The exhaust gas 'blows-down' under its own pressure and before the exhaust valve or ports close, pressurised air blows out a large part of the remaining exhaust gas and leaves the cylinder largely filled with fresh air, which is compressed after the exhaust valve closes at 70° after bottom dead-centre (abdc). This process is known as scavenging.

Scavenging can be symmetrical or unsymmetrical. In the least expensive form of two-stroke engines both inlet and exhaust are controlled by ports near the bottom of the cylinder and this inevitably produces a symmetrical timing. More efficient scavenging can be obtained if the exhaust valve closes before the inlet as this allows a degree of supercharge from the pressurised air, Figure 5.4.

The pressurised air in the least expensive engines is supplied from the crankcase where it is pressurised in the crankcase by the downward motion of the underside of the piston, Figure 5.5. Several methods of port scavenging are illustrated in Figure 5.6 and Figure 5.7, the latter a design of Professor G. Blair of Queen's University, Belfast. An efficient method of scavenging is known as the uniflow method. Figure 5.8 illustrates three examples.

With crankcase scavenging a very small amount of oil has to be added to the petrol to lubricate the piston and cylinder. However, in place of crankcase scavenging a positive displacement type blower (e.g.

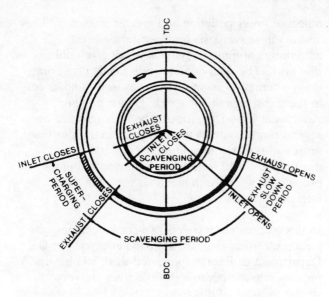

Figure 5.4 Timing Diagrams. Inner Diagram, Symmetrical Scavenge. Outer Diagram, Unsymmetrical Scavenge

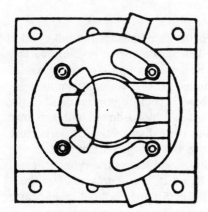

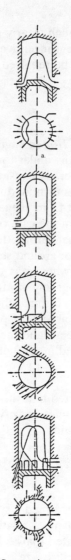

Figure 5.6 Methods of Scavenging. a. Cross Scavenging; b. Loop Scavenging, M.A.N. Type; c. Loop Scavenging, Schnuerle Type; d. Loop Scavenging, Curtis Type

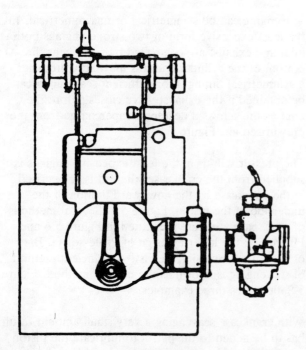

Figure 5.5 Sectioned Drawing of the OUB 30 Engine

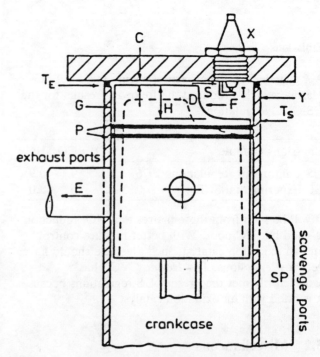

Figure 5.7 QUB Design of Deflector Piston and Combustion Chamber

98

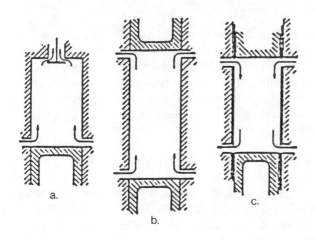

Figure 5.8 Uniflow Scavenging. a. Poppet Valve; b. Opposed Piston; c. Sleeve Valve

Roots type blower) driven by the engine may be used. This system has the advantage of permitting the use of a normal crankcase, as in a four stroke, with conventional lubrication.

Emission Characteristics

Four types of two-stroke engines are considered here:

(1) Conventional Mixture Scavenged Two-Stroke Engines (crankcase compression, spark ignition, two-stroke cycle engines scavenged with a mixture of air and fuel) are employed widely in motorcycle, utility and outboard marine applications. This engine type in comparison with the equivalent four-stroke cycle engine operating under similar conditions generally produces exhaust emissions with higher levels of HC and particulates, lower levels of NO_x, and similar levels of CO.

Published data [34] for a 350 cc road-going, sports motorcycle shows the exhaust HC to vary between 2500 and 9000 ppm (NDIR), while test results [35] for a 200 cc motor scooter show HC emissions in the range 1500 to 6000 ppm (NDIR). Comparisons [36] of small samples of utility and motorcycle engines show the HC emissions of mixture scavenged two-stroke engines to be between five and ten times greater than those for four-stroke engines.

At high loads the major factor in producing the high hydrocarbon emissions from mixture scavenged two-stroke engines is the loss of fresh charge directly to the exhaust system during the scavenging and trapping phases [34, 37]. At light loads irregular combustion contributes significantly to hydrocarbon emissions [34].

The gaseous hydrocarbon emissions are not affected by the oil/fuel ratio of the total loss lubrication systems employed in crankcase compression, two-stroke engines. However, particulate emissions do vary directly with oil/fuel ratio [38].

NO_x emissions from two-stroke engines are about half those from comparable four-stroke engines [36]. The lower NO_x emissions of the two-stroke engine can be attributed to the generally incomplete scavenging of combustion products from the engine cylinder even at full load. The retained combustion products provide inherent 'exhaust gas recirculation' and consequently reduced peak cycle temperatures and lower NO_x emissions. Published figures [39] for a small twin cylinder, mixture scavenged, two-stroke, automotive engine without emission control equipment show nitrogen oxide (NDIR) concentrations of between 60 and 800 ppm over the speed and load range.

(2) Stratified Charge Two-Stroke Engines, of which a number of experimental, mixture scavenged models have been built. Their inlet and scavenge porting systems are designed to give a stratified scavenging flow [40, 41]. The scavenging flow is divided into separate fuel-rich and fuel-lean mixture streams which are directed by the scavenge ports so as to reduce the total amount of fuel short circuited to the exhaust. These engines show significant improvements in thermal efficiency and exhaust hydrocarbon emissions. Hydrocarbon emissions of between 500 and 4000 ppm NDIR have been demonstrated [41]. These figures are still very much higher than those for a four-stroke engine.

(3) Direct-Injection Two-Stroke Engines are based on the theory that the elimination of fuel shortcircuiting can be achieved by injecting fuel directly into the engine cylinder after exhaust port closure. However, in practice it has proved necessary to inject during the open cycle in order to achieve sufficient mixing of air and fuel before ignition. Nevertheless, a study on a twin cylinder automotive engine [39] showed that by applying direct fuel injection exhaust hydrocarbons could be reduced to 230–2500 ppm (NDIR) for a mixture scavenged engine.

The two-stroke spark ignition engine is currently being examined as a possible future automotive powerplant by a number of manufacturers. Some claims [42] suggest that a multicylinder two-stroke engine with a unique air-blast direct injection system can produce specific fuel consumption and emission figures which are competitive with the best four-stroke engine valves but with reduced engine weight and packaging volume; though promising, the claims have not yet been conclusively demonstrated.

Current research is directed at applying improved fuel injection technology in order to achieve better fuel atomisation and improved combustion with later injection timings.

(4) Two-Stroke Diesel Engines are large low speed two-stroke diesel units widely used as marine main

propulsion units. These engine types probably have the highest thermal efficiency of any engine type in production. They do not exhibit the fuel short circuiting and associated high hydrocarbon emission characteristics of the mixture scavenged, two-stroke, spark-ignition engine.

Two-stroke diesel truck engines which meet the relevant emission legislation are in production in the USA.

ROTARY ENGINES

Many rotary engines have been and are still being proposed in the search for smaller, lighter and smoother power units. However, only one of these has reached the production stage. This is the configuration first proposed a century ago and later refined by Felix Wankel. The *Wankel* engine, shown diagrammatically in Figure 5.9, consists of an approximately triangular rotor mounted on and geared to a crankshaft, the centre of the rotor being eccentric to the crank centre. In rotating, the rotor turns the crank while the rotor tips describe an epitrochoidal path, a waisted ellipse. The engine casing or cylinder is machined to this form, which results in the rotor alternately compressing and expanding the working fluid and it is this feature together with its comparative simplicity which makes the power unit smaller, lighter and more smooth-running than the conventional reciprocating engine. The engine can be spark-ignition or diesel, but the only successful unit operates on the 4 stroke spark ignition cycle.

In its development, the first major problem to be solved was the efficiency and life of the sealing strips at the rotor tips, the rotor side seals giving no trouble. Much work was necessary before an acceptable effective life was obtained. This has been achieved although the seal life can never approach the life now being obtained with conventional piston rings.

The fundamental problem remains of the effect of the combustion chamber shape on the efficiency of combustion. Not only is the combustion chamber long and narrow, moving on its longer axis, but it involves a throat or narrowing at one point in the casing as can be seen in the diagram. Because of this, it is exceedingly difficult to produce a sufficiently homogeneous mixture in the combustion chamber and to ensure travel of the flame front to all parts of it in the time available. As a result the engine is substantially worse for emissions than the conventional piston engine, NO_x being about the same, CO worse and HC very much worse.

This situation has been improved but not eliminated by the use of both side and radial inlet porting and by multiple ignition systems, all adding to the complexity and losing many advantages of the engine.

The engine was developed in the 1960's by NSU and Fichtel and Sachs in Germany and taken up by Mazda in Japan who have continued the development. Other organisations such as *GM*, *Ford* and *Citroen* have done work on the engine but withdrawn when they have been able to evaluate the potential properly.

Early production cars used thermal reactors to reduce the HC. However, today three-way catalysts are required to meet regulations and the cost and requirements will be as with the conventional engines except that the catalysts will be required with all sizes of engine.

It is not expected that there will be any increase in usage of this type of engine.

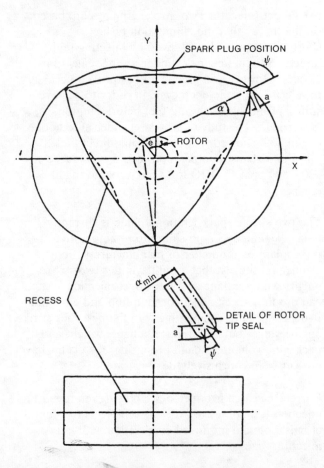

Figure 5.9 Combustion Chamber for Wankel Engine. a, Radius of Seal, 2ψ, Contact Angle of Seal, a, Crank angle, e, Eccentricities

GAS TURBINES

The potential advantage for the gas turbine lies chiefly in its light weight compared with diesel engines for large commercial vehicles. It is difficult to justify its use for passenger vehicles for either weight or volume. Experience has shown that the type of gas turbine suitable for vehicles is the so-called free turbine type in which one turbine drives the compressor and the free turbine drives the wheels of the vehicle through a suitable transmission. This system has an advantage over the conventional truck power unit in that the free power turbine gives a 3 to 1 increase of torque as the speed is reduced from full speed to stall, thereby reducing the requirements of the transmission. A two or three speed gearbox, with reverse, is required but this compares favourably with the 10 speed gearbox of many commercial vehicles.

The *Rover* vehicle gas turbine is shown diagrammatically in Figure 5.10 and a *BL*-designed one, sectioned in Figure 5.11. Air passes through the compressor and then through a heat exchanger that increases its temperature from heat derived from the exhaust gases. It then enters into the combustion chamber where it is mixed with fuel to give a well controlled flame. The hot gases produced are diluted by further air to a temperature the turbine blades can accept (about 1000°C) and after passing through the turbines are released to the atmosphere through the exhaust side of the heat exchanger.

It will be appreciated that the level of pollution will depend on the mode of combustion in the combustion chamber. Unlike the internal combustion engine the combustion is continuous and lends itself to a degree of control that is not possible in an IC engine.

The combustion chamber has three zones, the primary zone which burns at an air/fuel ratio near to stoichiometric, a secondary zone where excess air is fed in to give more complete combustion and an overall weak mixture and finally, as mentioned above, a dilution zone of tertiary air to reduce the temperature of the gases.

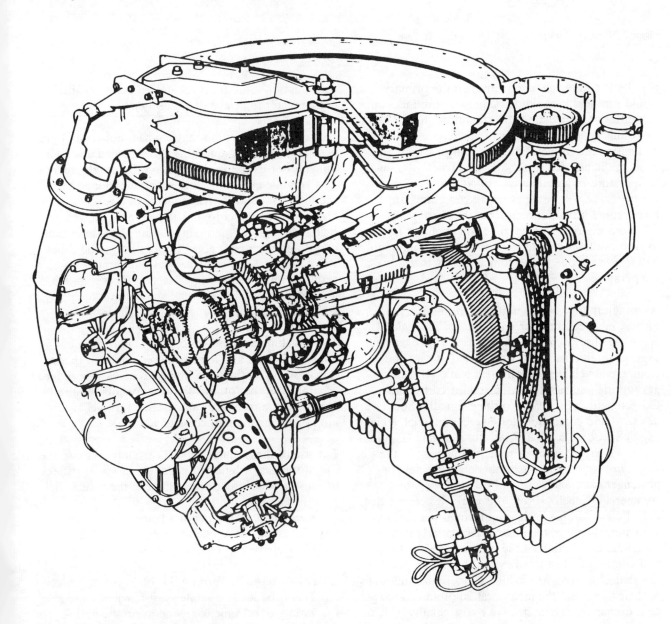

Figure 5.10 The Rover Vehicle Gas Turbine

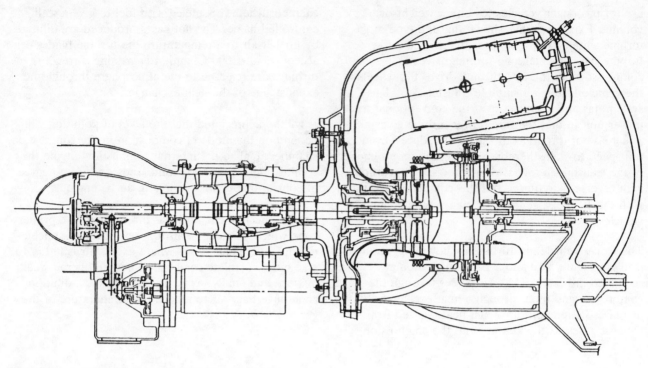

Figure 5.11 A BL Design for a 500 HP Vehicle Turbine

If pollution were not a concern then the primary zone would run near stoichiometric but this produces large quantities of nitric oxide. It is therefore desirable to make this zone somewhat richer to reduce the NO by limiting the oxygen. The secondary air then quickly cools the flame to a weak A/F ratio where the lower temperature reduces the rate of NO production. Clearly there will be areas where conditions are favourable for NO formation but the presence of some CO from the rich primary zone will compete for the oxygen available, the tertiary air will give minimum NO formation due to the comparatively low temperature.

As an alternative power unit the automotive gas turbine has been thoroughly evaluated in England, Germany, Italy and the US. In the UK, *Rover* built experimental units which were tested in cars and commercial vehicles. *Austin* built a 100 HP vehicle gas turbine and subsequently, sponsored by the Government, a 30 HP car turbine. Finally they designed and component tested a commercial vehicle engine based on a 250 HP industrial engine (Figure 5.11), which was in small scale production. In the US, *Chrysler* produced experimental turbines for passenger cars, and *Ford* and *GM* turbines for commercial vehicles. From this work, it became clear that the gas turbine in this form would not replace the conventional piston engine, particularly in the passenger car. Although the turbine can match the fuel consumption of the diesel at the design point, this demands using exotic high temperature materials at their limits and the incorporation of an expensive heat exchanger. These involve a cost penalty which is prohibitive. In addition, the increase in fuel consumption away from the design point is substantially greater than with the diesel engine.

However, this situation may change. Substantial programmes are going on in the USA, Japan and to a much lesser extent in the UK, on the development of new ceramic materials from which it is hoped that the high temperature components of a turbine will be able to be manufactured. These ceramics have a higher strength at high temperature than the nickel-based nimonic materials at present used, and this will permit the use of higher gas temperatures with a resulting improvement in efficiency. These materials are lighter than nimonics and should ultimately be cheaper.

The developments are concerned mainly in eliminating brittleness and improving resistance to thermal shock, while retaining an ability to be cast or sintered accurately in the form of the components required. Small turbine rotors for automotive turbo-chargers are now being made successfully.

The use of these materials will certainly increase the use of the gas turbine but is unlikely to be sufficient to cause a widespread application in the vehicle field.

In recent years, three major advanced gas turbine (AGT) systems have been financed in the USA by the US Department of Energy [43]:

(1) An *AiResearch/Ford* (AGT 101) system project. This is basically a single shaft machine, and a variety of ceramic components are being evaluated.

(2) *Detroit Diesel Allison* (AGT 100) (Division of GM)
This project provides for a target to meet the US Federal Emission Standards, and if successful would also meet the proposed Luxembourg Agreement figures. This is a two-shaft machine, and the projected fuel economy would be superior to its gasoline reciprocating counterpart.

(3) *Chrysler Williams* (AGT 102) Powertrain System Project
This is a single shaft machine with a complex continuously variable transmission incorporating a Van Doorne metal belt concept in a planetary gear system.

STIRLING CYCLE ENGINES

The Stirling cycle was one of the earliest proposed for a power unit but it has not been developed until relatively recently primarily because of lack of suitable materials. It has, however, potential advantages of very high efficiency, silence and low emissions against which must be weighed complexity, size and cost.

The major difference between the conventional internal combustion engines and the Stirling is that combustion in the latter is continuous and external to the power cylinder. It thus has the advantage, similar to the gas turbine, that combustion can be controlled better to give low emissions. The heat from combustion is passed, either directly from the combustion gases or via a heat pipe, into a heat exchanger from which it is transferred to the working fluid which is in a separate and sealed power circuit.

Two versions of the engine are possible, one with hydrogen or helium, usually the latter, in the power circuit and the other with air as the working fluid. Because of the thermodynamic properties of the working fluids, the helium engine gives lower losses and the theoretical efficiency is some 5 to 7% better than the diesel engine. The air engine gives an efficiency similar to or rather lower than the diesel engine.

The pistons of a Stirling engine are arranged in pairs, Figure 5.12, such that a 'cold' or displaced piston transfers working fluid through a heat exchanger where it picks up heat into a 'hot' or working cylinder where the heat is converted to work. The working fluid is then exhausted into another 'cold' space through a regenerator to store heat and then a cooler, from which the cycle starts again.

Control of power output is obtained by adjusting the mass and therefore the pressure of the working fluid actually in the circuit. This involves a complicated

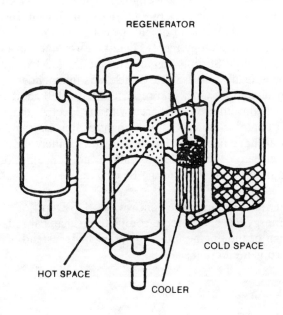

.95 PRINCIPLE CYCLE CONFIGURATION: WORKING GAS IN ONE OF THE CYCLES SHADED.

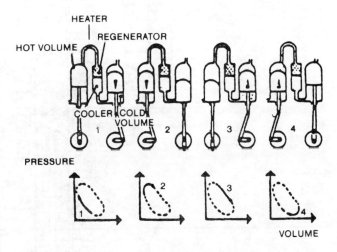

THE FOUR STROKES OF ONE WORKING CYCLE IN A DOUBLE-ACTING STIRLING ENGINE. NOTE THAT THE FOUR STROKES DO NOT DIRECTLY CORRESPOND TO THE FOUR PHASES OF THE IDEAL STIRLING PROCESS. CYCLE PHASING AND UNAVOIDABLE THERMODYNAMIC LOSSES RESULT IN A 'ROUNDED OFF' PV-DIAGRAM: (1) COMPRESSION AND HEATING OF WORKING GAS CAUSING RAPID PRESSURE INCREASE; (2) EXPANSION AT HIGH PRESSURE (WORKING STROKE); (3) EXPANSION AND COOLING OF WORKING GAS CAUSING RAPID PRESSURE DROP; (4) COMPRESSION AT LOW PRESSURE (COMPRESSION STROKE). ARROWS CORRESPOND TO A CLOCKWISE CRANKSHAFT ROTATION.

Figure 5.12 General Arrangement and Operating Principle of a Stirling Engine

system of valves, accumulators, etc., in order to get the rapid response necessary for vehicle applications.

There are many possible arrangements for a Stirling cycle engine but the most developed form is the Swedish *United Stirling* engine, Figure 5.13, which is based on earlier work by *Philips* of Eindhoven. In 1978 the US Department of Energy commissioned Mechanical Technology Inc. (MTI) to improve and evaluate the United Stirling engine. This engine (53 kW) in an American Motors Spirit vehicle was independently evaluated by *GM*, who reported that it fell short in fuel consumption compared with the Fleet Average for SI engines. However, in the light of this assessment MTI have improved the design with an increase of power to 64 kW and now claim a fuel consumption marginally superior to the SI engine.

In view of the promise shown, the US Dept. of Energy has sponsored a complete mechanical redesign of the engine using one crankshaft instead of the two geared crankshafts.

The high cost and relatively large size of such an engine arrangement must be obvious. Development has been held back however by some very difficult problems of which the performance and life of the heat exchangers and the life of the crosshead seals have been the major. Progress, however, has now been made in solving both of these.

RANKINE CYCLE ENGINE

In the early days of vehicle development, there was great interest in the use of the reciprocating steam engine (and steam turbine) because its torque output characteristic is much more suitable for traction purposes than the gasoline or diesel engine. For many years they were used for the larger commercial vehicles as well as on railways.

However, a basic maximum thermal efficiency of 8–10% as compared to the 25–35% efficiencies of

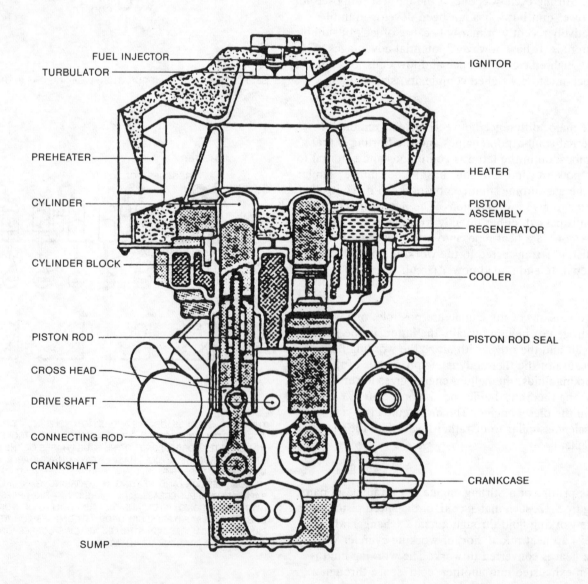

Figure 5.13 The United Stirling Engine

104

gasoline and diesel engines together with their greater bulk led to them being supplanted by the latter.

The main reason for low efficiency is due to the impracticability, in a mobile unit, of generating high enough 'gas temperatures'. A high pressure and high degree of superheating, possible in stationary applications are impracticable in vehicles. Complication and bulk is inevitable as a boiler and large condenser are required in addition to an engine. Indeed such schemes have some of the complications of the Stirling engine without the high efficiency of the latter.

Pollutants would theoretically be lower than in gasoline and diesel engines as continuous combustion is more easily controlled. The situation would be similar to the gas turbine and Stirling engine which are discussed elsewhere in this report. Thus interest was revived as interest in emissions increased, but there is no practical way of improving the fundamentally low efficiency sufficiently.

ADIABATIC AND COMPOUND DIESEL ENGINES

While much work continues on combustion and inspection systems in attempts to improve power, consumption and emissions, only the design of engines incorporating unit fuel injectors is likely to produce more than marginal improvements and even in this case they will be limited. More drastic changes are necessary.

The first of these, which is receiving attention mainly in the USA, Japan and Germany is the use of ceramic materials to insulate the combustion volume and thereby reduce the heat losses to the cooling water. This development has been christened the 'adiabatic' engine although it is not truly so. It gives a small improvement in power and specific fuel consumption (1.5–2.0% in efficiency) but its main advantages are in installation volume and weight and reduction in parasitic cooling losses.

There are no hard data on the effect on emissions but it can be expected that emissions would be reduced except in the case of NO_x which must be increased.

The major work in this field has been by *Cummins* [44] in the USA, sponsored by the US Army. Both lithium-aluminium silicate (LAS) and hot-pressed silicon nitride (HPSN) have been evaluated in arrangements as shown in Figure 5.14. LAS gives good insulation with poor high temperature strength and HPSN the reverse. It is understood that this programme is continuing in conjunction with ceramic

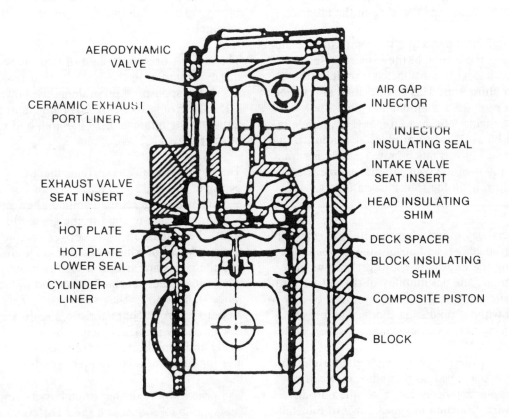

Figure 5.14 Cross Section of Cummins Basic Adiabatic (Insulated) Diesel Engine

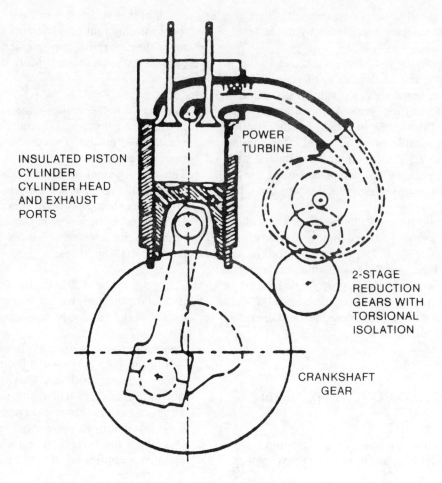

Figure 5.15 Total Energy Recovery Via Cummins Adiabatic Turbocompound Engine

developments to obtain a better balance of characteristics, with *NGK* in Japan leading in the latter.

The other effect of insulating the combustion chamber is to increase the energy in the exhaust gases and it is therefore logical to introduce a turbine in the exhaust to utilise this. This can be done in a variety of ways to produce a compound engine. The compound engine introduces a major advantage in that it can be arranged to produce an output torque characteristic which is much more suitable for traction purposes than the conventional engine which depends entirely on the transmission for its suitability.

The simplest way of introducing a turbine is to apply a larger turbocharger and absorb the power in compressing the intake air. However, this would merely increase the unsuitability of the output torque characteristic and would not allow enough energy to be utilised without producing combustion pressures which could not be accommodated.

It is necessary, therefore, to couple the turbine to the crankshaft in some way so that advantage can be taken of its characteristic that its torque output increases with reduction in speed down to the stall point. This can be done directly with a simple gear train, or by coupling it through a continuously variable ratio gear box (CVT) which allows better control of the balance in output between the engine and turbine.

The major work on the first of these is being carried out by *Cummins* in the USA in an extension of the US Army sponsored programme as shown in Figure 5.15. Their objective here is an engine of 500 HP with a fuel consumption of 0.28 lb/bhp h and an emission target (NO_x+HC) of 5 g/bhp h.

The more sophisticated arrangement using a CVT has been put forward by Wallace [45] of Bath University as part of his differential compound engine, Figure 5.16. In addition to the turbine geared to the output shaft, the crankshaft power drives an epicyclic gear, the sun wheel of which drives an inlet air compressor and the ratios are arranged such that the maximum inlet boost occurs at the bottom end of the speed range, falling off as this is increased. This produces an output torque characteristic, Figure 5.17, which is very nearly ideal. On first sight this scheme appears costly and complicated but this is not so as it eliminates the need for a costly main gearbox.

The practicability of this scheme was demonstrated 25 years ago by *Perkins* with their Differential Diesel Engine (DDE) which was extensively road tested, but was not pursued further.

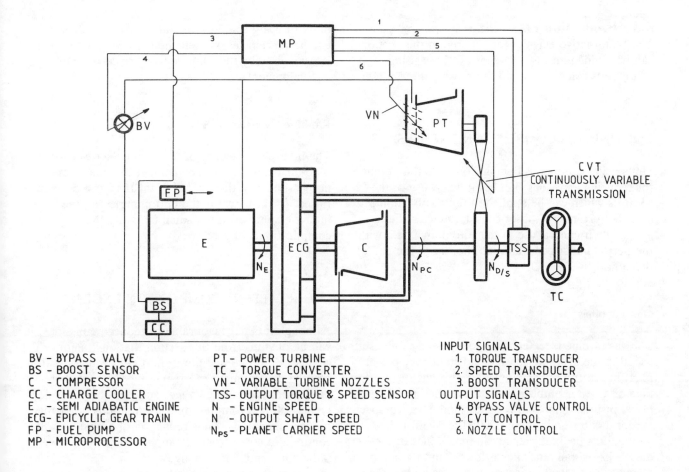

BV – BYPASS VALVE
BS – BOOST SENSOR
C – COMPRESSOR
CC – CHARGE COOLER
E – SEMI ADIABATIC ENGINE
ECG– EPICYCLIC GEAR TRAIN
FP – FUEL PUMP
MP – MICROPROCESSOR

PT – POWER TURBINE
TC – TORQUE CONVERTER
VN – VARIABLE TURBINE NOZZLES
TSS– OUTPUT TORQUE & SPEED SENSOR
N – ENGINE SPEED
N – OUTPUT SHAFT SPEED
N_{PS} – PLANET CARRIER SPEED

INPUT SIGNALS
1. TORQUE TRANSDUCER
2. SPEED TRANSDUCER
3. BOOST TRANSDUCER
OUTPUT SIGNALS
4. BYPASS VALVE CONTROL
5. CVT CONTROL
6. NOZZLE CONTROL

Figure 5.16 The Wallace Differential Compound Engine

This general line of development will lead to some improvement in engine specific weight and specific volume and to significant improvements in fuel consumption, probably more in terms of ton-miles per gallon than in lb/bhp h. The changes likely in the emission pattern are less clear, although it would be expected that emissions would decrease except for NO_x which would increase and require remedial action.

The road testing of the *Perkins* DDE, however, drew attention to two factors which may be beneficial in reducing total emission production. Firstly is the fact that on a gradient the compound engine speed is reduced, whereas in the conventional engine the engine speed is increased by running in an intermediate gear. Secondly, the elimination of gear changing has a substantial effect in improving journey time particularly on a hilly route, the implication of this being that the engine accumulates an appreciable time in a transient condition. Both these circumstances are likely to militate against good emission control in the conventional engine and the compound engine eliminates them.

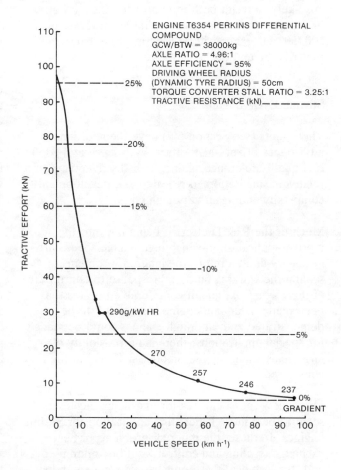

Figure 5.17 38 Ton Truck Tractive Effort Characteristics

107

5.3 Production Possibilities

The 'Alternative Engines' considered in this section can be divided into two categories, those using continuous combustion and those using intermittent.

CONTINUOUS COMBUSTION ENGINES

Units using continuous combustion have the potential for reducing emissions significantly because of the controls which can be applied in the combustion process. Unfortunately, other considerations militate against their general use in the automotive field within the period under consideration, i.e. up to the end of the century.

Gas Turbines

The gas turbine (see Section 5.2) is already well established in the aviation, marine and stationary power plant fields. In the vehicle field it also has the advantage of a better output torque curve than the conventional piston engine and is lighter and more compact except in sizes under about 200 HP. Against this, although its peak efficiency can match that of the diesel, its overall efficiency is lower, and its rate of response to control worse. However its major disadvantage at this time is cost. Nevertheless, component development programmes, particularly in ceramics, are likely to result both in improved efficiency and in reduced cost. It is possible that gas turbines in the 300–500 HP range will be used in heavy trucks towards the end of the century.

Stirling Cycle Engines

These units (see Section 5.2) have the additional advantages of very high efficiency, silence and good fuel quality tolerance against their disadvantages of a relatively small effort in research and development, complexity and again very high costs.

Much of the R & D effort is being put into the automotive application in which high engine speeds are necessary and therefore the inherent problems of sealing the working fluid and response to control are at their worst. Although considerable progress has been made, adequate engine life has still to be demonstrated and, although engines could be made for special applications, there is no possibility of large scale use in the automotive field within the time considered.

Large slow-speed engines for marine applications seem to be a more logical first application for Stirling engines. In these, the inherent problems of heat exchange, sealing and control would be minimised and, for instance, these units would offer the possibility of using coal as a fuel at higher than diesel engine efficiencies without the use of a gasifier. This is recognised by the Chinese who are carrying out such a development.

Rankine Cycle Engines

The inherent efficiency of these units (see Section 5.2) is so far below that of any competitors that there is no possibility of their application. If their efficiency is raised, for instance by the bottoming cycle, the improvement is still too small to justify the complication.

INTERMITTENT COMBUSTION ENGINES

Although it is not favourable to reduced emissions, it will be obvious from this section that any major changes in the mobile power unit field will come from onward development of existing types rather than the introduction of new types and that these developments will be prompted by considerations other than reduced emissions. Thus reduction of emissions will continue to come from the application of palliatives rather than any basic changes.

Stratified Charge Engines

These engines (see Section 5.1) comprise different approaches to the lean-burn concept with the advantages this gives in relatively low NO_x levels as well as a gain in efficiency. These are offset by the additional complexity and cost of the fuel injection equipment and the need for an oxidising catalyst to reduce HC emissions. Nevertheless, there is a general consensus that this is the most promising approach to the further development of the gasoline SI engine, and this view is supported by the fact that the *Honda* CVCC is in small scale production and the *Texaco* TCCS has been given extended trials in a parcel delivery service in the US.

Two Stroke Engines

Except in small sizes, two stroke engines (see Section 5.2) have problems with the cooling of the cylinder-head or combustion zone in the case of opposed-piston engines. This usually prevents full advantage being taken of the power capacity of the two stroke cycle, and application of turbocharging to four stroke cycle engines has destroyed the advantages of the two stroke in specific volume and specific weight.

On emissions, the inherent advantage of the two stroke in low NO_x emissions is balanced by the higher HC emissions arising both from the lubrication system and from fuel carry-over to the exhaust. The latter applies to the SI two stroke and can be eliminated by direct cylinder injection but this introduces limitations to the combustion and a substantial increase in cost. The direct cylinder injection in conjunction with a supercharger also gives the possibility of the cylinder cooling limitation being eased by using excess air but again with the penalties of complication and cost.

The existing market for SI two stroke engines in the small sizes, principally in mopeds and motor cycles, will continue and the improvements now possible will bring some small extension of this, although no major general change. In the case of the larger two stroke diesel, significant numbers of these have been produced in the UK, Germany and particularly by *GM* in the USA. However, problems primarily on cooling and engine life have led to these being dropped and it does not appear likely that this type of engine will reappear in its present form.

However, the 'adiabatic' engine avoids the major problem of cylinder cooling and when this type becomes established, quite a strong case can be made for an 'adiabatic' diesel two stroke unit as the gas producer for a compound engine – but beyond the timescale of this report.

Rotary Engines

Of the large number of possible rotary engines (see Section 5.2), only the Wankel engine has reached the stage of small-scale production and the advantages of this type do not appear likely to justify its wider adoption. For emissions, the shape of the combustion chamber makes it inherently worse than the conventional engine and this can only be corrected if means can be found to apply stratified charge principles to this engine.

Adiabatic and Compound Engines

The adiabatic engine (see Section 5.2) is a natural development from the conventional diesel engine. The objective of the US Army who is sponsoring much of this work is simpler installations in military vehicles such as tanks but the engine is also an intermediate step towards the compound engine.

While the main immediate purpose of compound engine development is increased efficiency and reduced fuel consumption, there is increasing understanding of the importance of the output torque curve obtainable from this type of engine and this understanding will lead to increasing investment in this line of development.

Wallace's ideas on the differential compound engine must be taken as the ultimate target for this line of development. There is, however, the possibility of development of the simple differentially supercharged diesel engine shown to be practicable by *Perkins* many years ago as an intermediate stage and it is likely that both lines will be followed. However, much work remains to be done before practical power units can be produced and it will certainly be ten years before any serious production is achieved.

Transmissions

The report has not dealt with the effect of transmissions on engine operating points or fuel economy, other than in the case of the differential compound engine. However, substantial developments are in hand on continuously variable transmissions which are likely to reach production in the reasonably near future both for cars and later for trucks. Electronic control of these will give the possibility of setting them for maximum economy or for minimum emissions and this can produce some significant variations to the current situations in both these elements although the actual extent of these has still to be determined.

6. Miscellaneous Emitters

6.1 Railways

6.2 Aircraft

6.3 Diesel Power Generation

6.4 Marine Engines

6. Miscellaneous Emitters

Emissions from mobile sources other than road transport are considered in this section. In the context of the acid deposition issue, where relatively long range transport of air pollutants up to 1000 km or more defines the spatial scale of the problem, emissions from these sources are relatively insignificant insofar as they generally comprise a small fraction of national emissions of the pollutants in question. This is not to say that local problems in the near vicinity of the sources considered here might not occur.

For the majority of sources considered in this section, with the possible exception of large civil aircraft engines, there is virtually no information available which has been obtained from measurements in, or is specific to, the UK. In these cases, where information is required, emission factors from the US EPA compilation [46] have been used.

6.1 Railways

There is virtually no information available on emission factors from UK railway locomotives. The largest source of acidic emissions from this sector will be diesel locomotives. Sulphur contents of diesel fuel currently in use in the UK will be 0.3–0.5% sulphur, so that given an annual consumption of diesel fuel in 1985 of 0.72 Mtonnes [47], the annual SO_2 emission from this source in the UK is 4.3–7.2 ktonnes or 0.1–0.2% of the national totals of all sources.

Emission factors appropriate to the UK for NO_x, HC and CO are not available. To obtain what must be regarded as approximate estimates, factors given by the US EPA [46] can be used. Approximate factors then are:

CO	19 g/kg of fuel
HC	13 g/kg of fuel
NO_x (as NO_2)	52 g/kg of fuel

Using these factors the national annual emissions of the three pollutants are respectively 14 kt, 9 kt and 37 kt respectively. These totals amount to 0.2%, 0.4% and 2% respectively of the national totals from all sources.

6.2 Aircraft

GENERAL

Aircraft engines are of two major categories: reciprocating piston and gas turbine.

In the piston engine, the basic element is the combustion chamber, or cylinder, in which mixtures of fuel and air are burned and from which energy is extracted by a piston and crank mechanism driving a propeller. The majority of aircraft piston engines have two or more cylinders and are generally classified according to their cylinder arrangement–either 'opposed' or 'radial'. Opposed engines are installed in most light or utility aircraft, and radial engines are used mainly in large transport aircraft. Almost no single row in-line or V-engines are used in current aircraft.

The gas turbine engine usually consists of a compressor, a combustion chamber and a turbine. Air entering the forward end of the engine is compressed and then heated by burning fuel in the combustion chamber. The major portion of the energy in the heated air stream is used for aircraft propulsion. Part of the energy is expended in driving the turbine, which in turn drives the compressor. Turbofan and turboprop (or turboshaft) engines use energy from the turbine for propulsion, and turbojet engines use only the expanding exhaust stream for propulsion. The terms 'propjet' and 'fanjet' are sometimes used for turboprop and turbofan, respectively.

In recent years the efficiencies and, correspondingly, emissions of carbon monoxide, hydrocarbons and smoke have been improved by the development of high-bypass ratio turbofan engines with low-emission combustion chambers. In the civil aviation area, all three major manufacturers, *Rolls-Royce*, *General Electric* and *Pratt and Whitney* are developing low emission combustors for turbine engines, and emissions in future years will decrease as these engines become more widely used. The most significant effect of the improved combustors is to reduce the emissions of CO and hydrocarbons in the idle mode (5% of full power) for an individual engine by up to a factor of about 5 for CO and up to an order of magnitude or more for hydrocarbons. Operation of the engine in idle/taxi mode represents the largest source of aircraft emissions of CO and hydrocarbons in the atmospheric boundary layer. The other major area of acidic emission is that of NO_x in the take off mode (100% full power); the improved combustor designs have very little effect on NO_x emissions.

THE LEGISLATIVE POSITION

The International Civil Aviation Organisation (ICAO) have promulgated standards regarding gaseous and particulate emissions from civil aircraft engines [48]. The gaseous emission standards for CO, HC and NO_x are to apply to engines whose date of manufacture is on or after 1 January 1986 (and whose rated thrust is greater than 26.7 kN). The gaseous emission levels are expressed in terms of

(i) D_p the mass of gaseous pollutant emitted in the reference Landing/Take-Off (LTO) cycle.

(ii) F_{oo} the rated output in kilo-Newtons (kN); the maximum power/thrust available for take-off under normal operating conditions at sea level.

(iii) π_{oo} the reference pressure ratio; the ratio of the mean total pressure at the last compressor discharge plane of the compressor to the mean total pressure at the compressor entry plane when the engine is developing take-off thrust rating in standard sea level conditions.

The standards are to be measured using defined procedures and are as follows:

HC: $D_p/F_{oo} = 19.6$
CO: $D_p/F_{oo} = 118$
NO_x: $D_p/F_{oo} = 40 + 2\pi_{oo}$

The particulate standard is expressed in terms of a Smoke Number, which is a dimensionless quantity measuring the smoke emission level based on the staining of a filter measured by a reflectance method, on the reference mass of exhaust gas sample.

The smoke standard or Regulatory Smoke Number is obtained from the reference method in a manner laid down by ICAO, and the standard, which shall apply to engines whose date of manufacture is on or after 1 January 1983 and is given by

Regulatory Smoke Number = $83.6 (F_{oo})^{-0.274}$
or a value of 50 whichever is lower.

The reference LTO cycle referred to in the regulations is given in Table 6.1.

Table 6.1

Operating Mode	Thrust Setting (as % of F_{oo})	Time in Operation Mode (mins)
Take-off	100	0.7
Climb	85	2.2
Approach	30	4.0
Taxi/ground idle	7	26.0

At the present time of writing (April 1987), the introduction in the UK of the ICAO standards is under discussion. They have been implemented in the USA only for CO and hydrocarbons.

EMISSION FACTORS

Three large compilations of emission factors for civil and military aircraft have been published by Pace [46], Sears [49] and the USEPA [50]. These compilations give emission factors and fuel consumption rates for a wide range of engines in the four standard modes, take-off, climb, approach and idle. However they are not necessarily representative of engines in current use in 1987 and these sources should only be used if no better data are available. In these references, total emissions per aircraft for a 'typical' LTO cycle are also quoted and an example is given in Table 6.2. These figures are based on a LTO cycle typical of a large congested metropolitan airport and give a broad indication of the magnitude of emissions from the aircraft listed. Emission calculations for particular airports or operations should use site specific data where this is available. The data presented in Table 6.2 and any emissions calculated from a LTO cycle, will typically refer to emissions occurring below the top of the boundary layer at 800–1,000 m. This will account for most of the emissions involved in the acid deposition processes but there will also be a contribution, primarily through the effect on ozone concentrations in the free troposphere (i.e. above the boundary layer and most weather systems) from emissions during higher level cruising.

Some site specific studies of UK airports have been carried out by WSL [51] and these suggest that civil aviation contributes a small fraction of the national UK totals, roughly 1% for CO and hydrocarbons and 0.5% for NO_x.

6.3 Diesel Power Generation

The material in this section draws heavily on a recent review paper by Hall and Wallin [52].

Stationary diesel engines are used for a variety of purposes, one of the commonest being electrical power generation. Capacities may vary from emergency standby generators of a few tens of kW up to continuously generating multiple units in power stations producing 100 MW or more in total. In spite of the wide range of sizes and types of engine used, they are all generically similar and use a limited range of fuels, the fuel oils ranging from gas oil up to heavy residual oil. As a result the pollutant content of the exhaust gases is quite characteristic.

The main gaseous pollutants are oxides of nitrogen and sulphur, carbon monoxide and the partially combusted hydrocarbons that produce, among other things, the characteristic 'diesel odour'. Particulate pollutants will be mainly carbon, some metals or their oxides and traces of polycyclic aromatic hydrocarbons (PAHs). The magnitude of the polluting effluent even from a small unit can be significant and some care is invariably required to ensure that it is safely and effectively discharged to the atmosphere.

Table 6.2 Emission Factors per Aircraft per Landing/Take-Off Cycle – Civil Aircraft

	Power Plant[a]			CO		NO_x[b]		Total HC[c]		SO_x[d]		Particulates	
	No	Mfg	Model-Series	lb	kg	lb	kg	lb	kg	lb	kg	lb	kg
Short, Medium, Long Range and Jumbo Jets													
BAC/Aerospatiale Concorde	4	RR	Olymp 593	847.00	384.00	91.00	41.0	246.00	112.00	14.10	6.40		
BAC 111-400	2	RR	Spey 511	103.36	46.88	15.04	6.82	72.42	32.85	1.70	0.77	1.46	0.66
Boeing 707-320B	4	P&W	JT3D-7	262.64	119.12	25.68	11.64	218.24	99.00	4.28	1.94	4.52	2.05
Boeing 727-200	3	P&W	JT8D-17	55.95	25.38	29.64	13.44	13.44	6.09	3.27	1.48	1.77	0.53
Boeing 737-200	2	P&W	JT8D-17	37.30	16.92	19.76	8.96	8.96	4.06	2.18	0.99	0.78	0.35
Boeing 747-200B	4	P&W	JT9D-7	259.64	117.76	83.24	37.76	96.92	43.96	7.16	3.25	5.20	2.36
Boeing 747-200B	4	P&W	JT9D-70	108.92	49.40	107.48	48.76	22.40	10.16	7.96	3.61	5.20	2.36
Boeing 747-200b	4	RR	RB211-524	66.76	30.28	124.90	56.65	10.00	4.54	7.52	3.41		
Lockheed L1011-200	3	RR	RB211-524	50.07	27.71	93.66	42.48	7.50	3.40	5.64	2.56		
Lockheed L1011-100	3	RR	RB211-22B	199.40	90.44	64.29	29.16	138.40	62.77	4.95	2.24		
McDonnell-Douglas DC8-63	4	P&W	JT3D-7	262.64	119.12	25.68	11.64	218.24	99.00	3.27	1.48	1.17	0.53
McDonnell-Douglas DC9-50	2	P&W	JT8D-17	37.30	16.92	19.76	8.96	8.96	4.06	2.18	0.99	0.78	0.35
McDonnell-Douglas DC10-30	3	GE	CF6-50C	116.88	53.01	49.59	22.17	47.10	21.36	4.98	2.26	0.21	0.10
Air Carrier Turboprops – Commuter, Feeder Line and Freighters													
Beech 99	2	PWC	PT6A-28	7.16	3.25	0.82	0.37	5.08	2.30	0.18	0.08		
GD/Convair 580	2	A11	501	24.38	11.06	21.66	9.82	9.82	4.45	0.92	0.42		
Dehavilland Twin Otter	2	PWC	PT6A-27	7.16	3.25	0.82	0.37	5.08	2.30	0.18	0.08		
Fairchild F27 and FH227	2	RR	R.Da.7	36.26	16.45	0.92	0.42	22.42	10.17	0.58	0.26		
Grumman Goose	2	PWC	PT6A-27	7.16	3.25	0.82	0.37	5.08	2.30	0.18	0.08		
Lockheed L188 Electra	4	A11	501	48.76	22.12	43.32	19.65	19.64	8.91	1.84	0.83		
Lockheed L100 Hercules	4	A11	501	48.76	22.12	43.32	19.65	19.64	8.91	1.84	0.83		
Swearingten Metro-2	2	GA	TPE 331-3	6.26	2.84	1.16	0.53	7.68	3.48	0.16	0.07	0.46	0.21
Business Jets													
Cessna Citation	2	P&W	JT15D-1	19.50	8.85	2.00	0.91	6.72	3.05	0.40	0.18		
Dassault Falcon 20	2	GE	CF700-2D	76.14	34.54	1.68	0.76	7.40	3.36	0.78	0.35		
Gates Learjet 24D	2	GE	CJ610-6	88.76	40.26	1.58	0.72	8.42	3.82	0.84	0.38		
Gates Learjet 35, 36	2	GE	TPE 731-2	11.26	5.11	3.74	1.58	3.74	1.70	0.92	0.42		
Rockwell International Shoreliner 75A	2	GE	CF 700	76.14	34.54	1.08	0.76	7.40	3.36	0.78	0.35		
Business Turboprops (EPA Class P2)													
Beech B99 Airliner	2	PWC	PT6A-27	7.16	3.25	0.82	0.37	5.08	2.30	0.18	0.08		
Dehavilland Twin Otter	2	PWC	PT6A-27	7.16	3.25	0.82	0.37	5.08	2.30	0.18	0.08		
Shorts Skyvan-3	2	GA	TPE-331-2	6.44	2.92	0.88	0.40	8.40	3.81	0.16	0.07	0.46	0.21
Swearingen Merlin IIIA	2	GA	TPE-331-3	6.28	2.85	1.15	0.522	7.71	3.50	0.16	0.07	0.46	0.21
General Aviation Piston (EPA Class P1)													
Cessna 150	1	Con	0-200	8.32	3.77	0.02	0.01	0.23	0.10	0.0	0.0		
Piper Warrior	1	Lyc	0-320	14.37	6.52	0.02	0.01	0.26	0.12	0.0	0.0		
Cessna Pressurised Skymaster	2	Con	TS10-360C	33.10	15.01	0.13	0.06	1.15	0.52	0.0	0.0		
Piper Navajo Chieftain	2	Lyc	T10-540	96.24	43.65	0.02	0.01	1.76	0.80	0.0	0.0		

a Abbreviations: A11–Detroit Diesel Allison Division of General Motors; Con–Teledyne/Continental; GA–Garrett AiResearch; GE–General Electric; Lyc–Avco/Lycoming; P&W–Pratt & Whitney; PWC–Pratt & Whitney Aircraft of Canada; RR–Rolls Royce
b Nitrogen oxides reported as NO_2
c Total hydrocarbons. Volatile organics, including unburned hydrocarbons and organic pyrolysis products.
d Sulphur oxides and sulphuric acid reported as SO_2

SULPHUR DIOXIDE (SO_2)

This pollutant is produced from combustion of sulphur in the fuel. Virtually all the sulphur is discharged in the exhaust gases as SO_2. Fuel sulphur contents will vary from about 0.3–0.5% for distillate fuels up to about 3.5% for residual fuel oils. The typical sulphur content for a blended medium oil for stationary engine use is around 2%. A small trace of sulphur trioxide (SO_3, a more toxic and corrosive pollutant than SO_2) is also produced; this is not normally significant unless the surfaces of the exhaust ducts and chimney are allowed to fall below the acid dewpoint (about 135°C) when condensation on to the walls of chimneys and ducts provides sites for the deposition of carbon and the formation of sulphuric acid. Acid and carbon agglomerate can be spasmodically discharged as acid smuts.
Concentrations of SO_2 in the exhaust gases will vary from about 90 ppm (0.3% S fuel) to about 1000 ppm (3.5% S fuel) (6 g/kg fuel to 70 g/kg fuel). By comparison, concentrations in boiler plant range from about 180 ppm (0.3% S fuel) to about 2100 ppm (3.5% S fuel) (6 g/kg fuel to 70 g/kg fuel). The rate of discharge per kg of fuel consumed is the same as from diesel exhaust but because of the lower exhaust gas volume the concentration of SO_2 in the stack is higher.

OXIDES OF NITROGEN (NO_x)

These are formed from oxygen and nitrogen present during the combustion process (there may be small amounts of nitrogen additionally present in the heavier fuels; this is also converted to nitrogen oxides). The amount of NO_x produced depends on the temperature, pressure and detailed features of the combustion. Since diesel combustion uses high temperatures and pressures, diesel exhaust concentrations of NO_x are relatively high. The major constituent is nitric oxide, NO, and levels vary typically from 400 to 700 ppm (12.5–22 g/kg fuel). By comparison, a conventional boiler plant would have flue gas NO concentrations around 150–200 ppm (2.1–3.2 g/kg fuel). The principal remaining constituent of the total emission of nitrogen oxides is nitrogen dioxide (NO_2, a more toxic pollutant than NO) and is typically from 5–10% of the NO concentration for diesel engine exhausts, thus maximum values are about 70 ppm (3.3 g/kg fuel). It is usually less than 5% in the flue gases from oil fired boilers, that is about 10 ppm maximum (0.23 g/kg fuel).

CARBON MONOXIDE (CO)

Carbon monoxide is a product of partial combustion. Concentrations in diesel exhaust flue gases are typically in the range 200–300 ppm (6–9 g/kg fuel). These levels are about four times those found in a conventional boiler, which are generally around 50 ppm (0.7 g/kg fuel).

TOTAL HYDROCARBONS (HC)

Levels are relatively low, because of the high efficiency of diesel combustion, around 25–40 ppm, expressed as C_3 (1.2–1.9 g/kg fuel). Levels in boiler flue gases are similar at 25–40 ppm (0.6–0.9 g/kg fuel).

DIESEL ODOUR

The characteristic diesel odour is produced mainly by hydrocarbon products of partial combustion together with the odour of nitrogen oxides. The 'burnt' character of the odour is associated with furans, alkylbenzaldehydes and other oxygenated organic compounds. The exhaust gases must typically be diluted about 5,000 times to ensure that the odour level is unobtrusive.

PARTICULATES

The particulate content of the exhaust is mainly carbon, but it may also contain low levels of Poly-Aromatic-Hydrocarbons (PAHs) together with metal oxides from metallic compounds present in the fuel (the most important being vanadium). Particulate concentrations in the exhaust gases will vary considerably with the fuel burned, from about 7.5 mg/m^3 with the lighter distillate fuels, an almost negligible level, up to 300 mg/m^3 with heavy residual oils (0.2 g/kg fuel–7 g/kg fuel).

The PAHs are composed of a large number of different compounds; over 250 have been identified in diesel emissions. Many of these compounds have only recently been identified and their toxicity and mutagenicity are still under investigation. Emission levels are very small, the maximum is unlikely to exceed 40 μg/kg fuel.

EMISSION ABATEMENT METHODS

It is possible to abate the emissions in three ways. Firstly by control of the fuel used. The use of the lighter fuels can reduce some of the pollutant emissions, notably particulates and sulphur oxides. They can also be useful in alleviating additional problems during start up and shut-down; in plant normally using heavy fuel a light fuel can be substituted during these periods. Secondly, the combustion process can be modified; this is of particular interest in the reduction of NO_x emissions. Finally, a variety of abatement systems can be applied to the exhaust gases to reduce some components of the emission. These latter two options are reviewed briefly below.

Modification of the Combustion Process

Altering the way in which fuel is burned in the engine can alter the level of pollutants emitted in the exhaust gases. This is of particular interest as a method of controlling NO_x emissions, the general rule being that any modification which reduces the peak combustion temperature and pressure in the engine reduces the level of NO_x emitted. Apart from changes to the engine design, which will not be considered here, it is possible to control the combustion by modifying the fuel or the combustion air. Two examples are discussed, water injection and exhaust gas recirculation.

The addition of water to the fuel reduces the combustion temperature and, thereby, the NO_x emission level. This method is used in gas turbines, but, so far, not significantly in diesel engines. Water contents up to 20% can be used, but the largest reduction in NO_x levels so far obtained is around 10%. This is a relatively small reduction for the effort involved and probably explains the limited interest in the method.

A more promising technique is exhaust gas recirculation (EGR). Because all the combustion air is not used in a single pass, it is possible to recirculate part of it back into the engine. The maximum is about 10%, but this is sufficient to affect the combustion in a favourable way. There are two distinct mechanisms which operate. Firstly, additional heat energy is absorbed during combustion and, secondly, the dilution of the charge reduces the flame speed. Peak combustion temperature and pressures are thereby reduced and consequent reductions of the level of NO_x of the order of 30% are attainable. Because the A/F ratio varies with load it is necessary to vary the amount of exhaust gas recirculated as a function of load and this must be done automatically. Simple systems can use a control valve but better results can be obtained from more complex arrangements using A/F sensors coupled with microprocessor control. Difficulties can arise with turbocharged engines where the inlet/exhaust pressure difference may be insufficient to drive the EGR and additional recirculating fans may be needed. The potential of EGR seems to be best realised at part-load conditions; at high loads there is less oxygen available in the recirculated air.

Abatement Methods

Abatement of emission may be by any one of three techniques:

(1) Catalytic oxidation units, heated by the exhaust gases themselves, will oxidise remaining products of partial combustion, mainly hydrocarbons and CO. Thus the principal effect is on the odorous hydrocarbon component of the emission, some of which is removed, typically 50% or more. There is no effect, of course, on already oxidised material such as SO_2 and there is a deleterious effect in further oxidising some NO (typically 5–10%) to NO_2, which is a more toxic pollutant. A further development of the catalytic oxidiser unit incorporates an arrangement of the catalyst which also acts as a filter to trap and then oxidise the carbonaceous particulate.

(2) Scrubbers can remove some proportion of the soluble pollutants, SO_2 and NO_x, and can also remove particulates. With a single stage scrubber up to 50% of SO_2 can be removed and 50% or more of NO_2. Nitric oxide, NO, has low solubility and is not removed in significant quantities by a single stage scrubber. A two stage unit is much more effective, about 90% removal can be obtained. Owing to the small size of diesel exhaust particulates, only a high energy venturi scrubber will be effective in removing them.

Scrubbers are not commonly used on diesel exhausts. They are difficult to use where there is no exhaust heat recovery and there are deleterious effects from excessively cooling the gases and loading them with water vapour. This can cause difficulties with high plume visibility in cold weather. Also effective dispersion in the atmosphere is reduced due to low plume buoyancy. There can, in addition, be unwanted generation of SO_3.

(3) Removal of particles by various means. There are two effective methods that could be used for removal of the small particles of 2 μm diameter which constitute the greater part of diesel exhaust particulates (typically 90% or more by mass). These are bag filters and electrostatic precipitators. Both require reduction of the gas temperature to below 300°C and are expensive to install and maintain. Inertial separators, such as cyclones or other swirl inducing devices, are only effective on the larger agglomerated particles which may be formed in the duct or heat recovery systems.

EMISSION LIMITS TO DIESEL DISCHARGES

Many countries have regulations limiting the pollution emissions from combustion plant in order to control large scale pollution problems. In most regulations stationary diesel sources are included with other combustion plant; Table 6.3 lists the emission limits for stationary sources using liquid fuel for a number of European Countries and the USA for sulphur dioxide, nitrogen oxides and particulates, the three major emitted pollutants. Also quoted are limits recently proposed for EEC member states for large combustion plant, they are currently still under discussion. All the limits are for concentrations of pollutant within the stack rather than emission per kg of fuel consumed and this gives rise to a problem in applying them to diesel discharges. They are mainly concerned with steam raising plant where combustion is at near stoichiometric air/fuel mixtures and the limits are referred to a flue gas oxygen content of about 3%. For diesel plant running typically at 100% excess air the flue gas volume will be roughly doubled, so that on a basis of emission of pollutants/kg fuel consumed, the concentrations quoted in the table for typical diesel discharges should be roughly doubled for comparison with the various emission limits. This point has not been clarified in most existing legislation.

It can be seen from Table 6.3 that in many cases the existing or proposed emission limits cannot be satisfied without either restricting the fuel to the lighter fractions and/or by the use of abatement equipment.

For particulates the emission limits for France and the UK are difficult to meet, without abatement, in the case of some residual fuels. The proposed limits in the EEC directive would require the use of distillate fuels or particle abatement where residual oils are used.

Only the FRG and the USA currently impose limits on SO_2 emissions. In both cases either SO_2 abatement or restriction to fuels of 0.5% and 0.7% sulphur respectively, as measured, would be required. The proposed EEC directive would require either a limitation to fuels of less than about 1% S content (based upon a reference of 3% O_2 in the flue gases) or the use of abatement.

Again only the FRG and the USA currently apply limits to NO_x emissions. The proposed EEC limit is lower, but of a similar order. These are achievable for diesel exhaust concentrations, as measured, possibly with some abatement (e.g. EGR). However, if based upon a 3% O_2 reference level, conforming to this requirement would be difficult and expensive with existing technology.

6.4 Marine Engines

There is virtually no information available on marine engine emissions in the UK and Europe and that which is available in the USA is relatively approximate. In terms of national acidic emissions, the marine sector is relatively unimportant – UK SO_2 emissions from these sources for example comprise about 0.3% of the national total.

Some emission factors for diesel vessels for a range of power outputs are given for CO, HC and NO_x in Table 6.4. Emissions of SO_2 are most conveniently calculated from the sulphur content of the fuel, which in the UK is generally 0.3–0.5% by weight. The data in Table 6.4 is from the US EPA [50].

Emission factors for pleasure craft are given in Table 6.5; they also originate from the US EPA [50]. Both sets of data should be regarded as approximate, particularly when applied to the UK or Europe.

Table 6.3 Typical Diesel Emissions Compared with Emission Limits for Combustion Plant

Pollutant	Size of Plant MW Thermal Input	Typical** Concentrations mg/Nm³ for Stationary Diesels	Emission Limit Values, mg/Nm³ at 3% O_2				
			EEC*	UK	France	USA	FRG
Particles	100	7.5¹–300²	50	115	100	350	–
Particles	100–300	7.5¹–300²	50	115	100	350	–
Particles	300	7.5¹–300²	50	115	100	350	–
SO_2	100	230¹–2600²	1700	–	–	920	650
SO_2	100–300	230¹–2600²	1700	–	–	920	
SO_2	300	230¹–2600²	400	–	–	920	150
NO	100	540–940	450			570	950
NO	100–300	540–940					
NO	300	540–940					220

* Proposed
** See text
1 For Distillate Fuels, 0.3% S
2 For Residual Fuel Oils, 3.5% S

Table 6.4 Diesel Vessel Emission Factors by Operating Mode

Horsepower	Mode	Emissions					
		CO		HC		NO$_x$ as (NO$_2$)	
		lb/10^3 US gallons	kg/10^3 litre	lb/10^3 US gallons	kg/10^3 litre	lb/10^3 US gallons	kg/10^3 litre
200	Idle	210.3	25.2	391.2	46.9	6.4	0.8
	Slow	145.4	17.4	103.2	12.4	207.8	25.0
	Cruise	126.3	15.1	170.2	20.4	422.9	50.7
	Full	142.1	17.0	60.0	7.2	255.0	30.6
300	Slow	59.0	7.1	56.7	6.8	337.5	40.4
	Cruise	47.3	5.7	51.1	6.1	389.3	46.7
	Full	58.5	7.0	21.0	2.5	275.1	33.0
500	Idle	282.5	33.8	118.1	14.1	99.4	11.9
	Cruise	99.7	11.9	44.5	5.3	338.6	40.6
	Full	84.2	10.1	22.8	2.7	269.2	32.3
600	Idle	171.7	20.6	68.0	8.2	307.1	36.8
	Slow	50.8	6.1	16.6	2.0	251.5	30.1
	Cruise	77.6	9.3	24.1	2.9	349.2	41.8
700	Idle	293.2	35.1	95.8	11.5	246.0	29.5
	Cruise	36.0	4.3	8.8	1.1	452.8	54.2
900	Idle	223.7	26.8	249.1	29.8	107.5	12.9
	2/3	62.2	7.5	16.8	2.0	167.2	20.0
	Cruise	80.9	9.7	17.1	2.1	360.0	43.1
1580	Slow	122.4	14.7	–	–	371.3	44.5
	Cruise	44.6	5.3	–	–	623.1	74.6
	Full	237.7	28.5	16.8	2.0	472.0	5.7
2500	Slow	59.8	7.2	22.6	2.7	419.6	50.3
	2/3	126.5	15.2	14.7	1.8	326.2	39.1
	Cruise	78.3	9.4	16.8	2.0	391.7	46.9
	Full	95.9	11.5	21.3	2.6	399.6	47.9
3600	Slow	148.5	17.8	60.0	7.2	377.0	44.0
	2/3	28.1	3.4	25.4	3.0	358.6	43.0
	Cruise	41.4	5.0	32.8	4.0	339.6	40.7
	Full	62.4	7.5	29.5	3.5	307.0	36.8

Table 6.5 Average Emission Factors for Inboard Pleasure Craft

Pollutant	Based on fuel consumption				Based on operating time			
	Diesel engine		Gasoline engine		Diesel engine		Gasoline engine	
	kg/10^3 l	lb/10^3 US gallons	kg/10^3 l	lb/10^3 US gallons	kg/hr	lb/hr	kg/hr	lb/hr
SO$_x$ as SO$_2$	3.2	27	0.77	6.4	–	–	0.008	0.019
CO	17	140	149	1240	–	–	1.69	3.73
HC	22	180	10.3	86	–	–	0.117	0.258
NO$_x$ as NO$_2$	41	340	15.7	131	–	–	0.179	0.394

7. In-Service Emissions Performance and Inspection

7.1 Durability of Control Systems In-Service

7.2 In-Service Inspection

7.3 Summary

7. In-Service Emissions Performance and Inspection

7.1 Durability of Control Systems In-Service

The in-use vehicle may not necessarily conform with the standard to which it was built. However, most vehicles may be tuned to the manufacturers' specifications and when tested in accord with the ECE R15 test schedule produce emissions which are within the prescribed limits. Some results for two sets of conventional engined cars are given in Figure 7.1 and Figure 7.2. Also, the effect of tuning is to reduce fuel consumption as is shown in Figure 7.3.

Although the two groups of small engined cars were fuelled with variable jet carburettors it is clear that the settings of the carburettors in the first group were unstable. The instability led to fuel enrichment across the engine operating range with consequent high CO and HC emissions and enhanced fuel consumptions. The effect of a major service, which returned the emissions related settings to the manufacturer's specifications, was to reduce CO, THC emissions and fuel consumptions, and increase NO_x emissions over the operating range of the vehicles.

The second group of cars at their first major service test point were much less in breach of the Conformity of Production (COP) standards than the first group of cars and hence the adjustments made during the service test procedures did not have a major effect on the calculated fuel consumptions.

However, the results show that for some cars it is beneficial to the owner and to the environment for regular inspection and adjustments to be made to simple engine parameters. In a study of 204 cars manufactured in 1971 and 1982 Williams and Everett [54] showed that 95 in-service cars which were fitted with variable jet carburettors were emitting CO and HC levels which were on average above the appropriate Type Approval (TA) level. For example in the case of non-tamper proofed carburettors the ECE R15 mean emissions levels for CO and THC were 206 and 120% of the TA standard. In the case of 109 cars fitted with fixed jet carburettors relatively high CO emissions were also observed; for example, the non-tamper proofed carburetted cars had mean CO emissions that were 151% of the TA standard. The HC and NO_x emissions were 101 and 45% of the

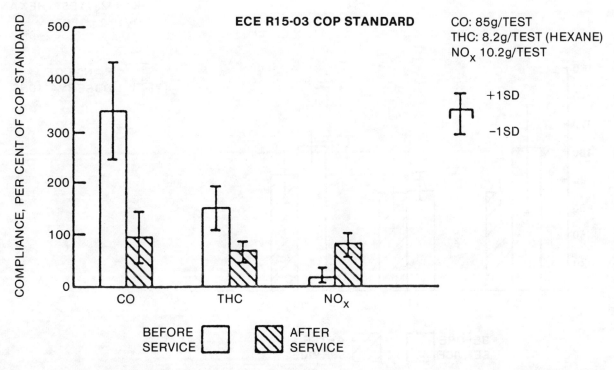

Figure 7.1 Comparison of Cold-Start ECE R15 Emissions from Four 1.0 Litre Cars Before and After Service Over an In-Service Life of Approximately 40,000 Miles

TA standard respectively. The effect of tuning was to reduce CO and THC emissions and increase NO$_x$ emissions; a small overall benefit (3%) in fuel consumption was observed.

In the same study the authors separated the then relatively new cars, i.e. those which were built to the ECE R15-03 regulations. Although there was an improved compliance with the standard it should be noted that the garage (forecourt-sales) cars had mean CO emissions 70% above the limit. It is of interest to note that the carbon monoxide content of the

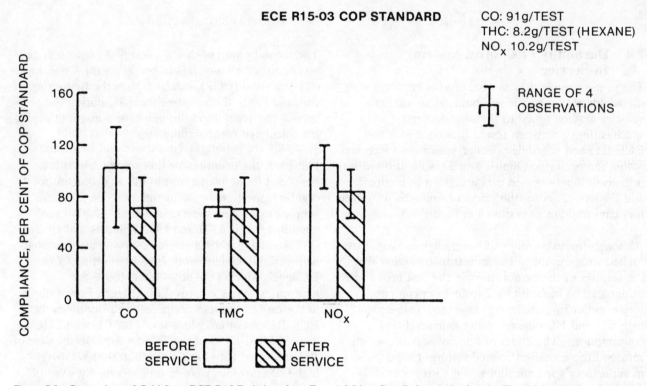

Figure 7.2 Comparison of Cold-Start ECE R15 Emissions from Four 1.3 Litre Cars Before and After the First Major Service at About 10,000 Miles

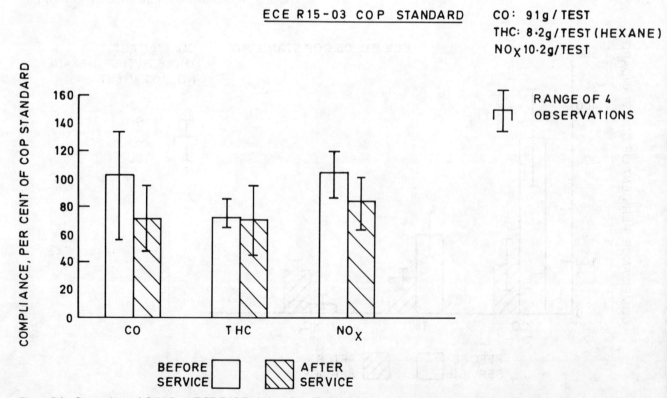

Figure 7.3 Comparison of Cold-Start ECE R15 Emissions from Four 1.3 Litre Cars Before and After the First Major Service at about 10,000 Miles

exhaust gases at idle were above the control limit (3–5%). Table 7.1 shows the levels of compliance with the standard of 41 cars sourced at garages, hire companies, privately owned and company cars.

Table 7.1 Comparison of Results of Cars from Different Sources for the 03 Approval Level

Source*	No. of Cars	Mean Emission Levels, % of Weighted Standard			
		CO	HC	NO_x	CO at Idle
H	12	104	84.6	67.2	94.2
P	7	149	93.1	56.7	176
G	13	170	103	55.9	189
C	9	125	79.4	69.3	120

*H = Hire; P = Private; G = Garage; C = Company

In the study of the emissions performance of two groups of cars from new it has been shown that the in-service vehicle may be a high emitter of carbon monoxide and hydrocarbons. The latter group of compounds contains precursors of photochemical oxidants and if it were considered necessary to reduce hydrocarbon emissions, effective control in service could be effective. However, it is important to note that the data for the first major services of the second group of cars indicated a relatively stable state of tune. Both groups of cars after routine major services were, on average, emitting regulated pollutants within the appropriate standard. Furthermore, improvements in engine fuelling and engine management techniques may be expected to improve the in-service emissions and fuel economy stability of new vehicle types.

In a study of 136 catalyst equipped cars operated by the Hessen Police [55] it has been shown that servicing can restore the low emissions performance of these vehicles. Plots of emissions of CO, HC and NO_x over 80,000 km are given in Figures 7.4, 7.5 and 7.6.

7.2 In-Service Inspection

INTRODUCTION

In-service inspection of motor vehicles is a procedure designed to ensure that emissions-related build standards are maintained in service and hence the objectives of emissions controls are not frustrated. In the US there have been various schemes on a state basis to implement inspection of vehicles. Also, the Environmental Protection Agency has carried out surveys of in-use vehicles in order to assess the effectiveness of the emissions control systems used by manufacturers, which have been certified to be effective over a service life of 50,000 miles.

In the European Community consideration has been given to inspection by France, FRG and the Netherlands. The approaches have diverged widely both with regard to the inspection procedure and to the limit levels enshrined in national legislation. Elsewhere, Austria, Japan, Switzerland and Sweden have introduced their systems. The European

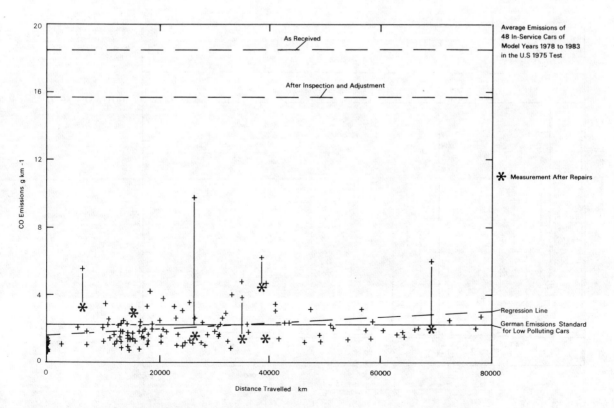

Figure 7.4 CO Emissions in the US 1975 Test Procedure: 136 Three-Way Catalyst Equipped Cars

Commission has a mandate to propose a Directive on in-service inspection for safety related items in which the Commission proposals include emissions inspection.

In the case of conventional engine technology options which are capable of meeting the ECE R15-04 regulations, carburettor tamper-proofing and regular maintenance have been considered to be adequate to ensure reasonable adherence to standards. However, the 'Luxembourg' agreement and the adoption of emissions standards which will require the use of catalysts (oxidation and three-way) for some cars and the fiscal incentives of the FRG for motorists to purchase 'low polluting' cars have emphasised the possible requirements of official in-service test procedures.

OFFICIAL TEST PROCEDURES

Official test procedures imply a number of decisions to be made by national legislation. For example, which pollutant(s) should be controlled, what are the limit values, what is the frequency of inspection required, what charge to the motorist can be made, should the test be additional to an existing (safety related) inspection? Historically, the most difficult decisions have been with regard to the test procedures which, in the US, were originally designed to be a surrogate for a full emissions test. These procedures were reviewed by Kueper and Porsche [56]. The general approach, however, is to measure the concentrations of carbon monoxide and hydrocarbons in the exhaust gases under engine idling conditions; in the case of catalyst equipped cars a loaded test on a simple chassis dynamometer may be required.

A general overview of international in-service emissions regulations is given in Table 7.2.

Generally, in the cases of European states the carbon monoxide control levels for idle were based on the Type II Test of ECE R15 (see Section 1, Table 1.1), i.e. 4.5% which was reduced to 3.5% in the '03' amendment. However manufacturers' tuning data for modern cars which have been homologated to ECE R15-04 tend to be for levels which may be considerably less than 3.5% of CO at idle. For example, the 1980/86 Ford Escort idle CO content is $1.5 \pm 0.5\%$ at an idle speed of $800+25$ rev/min in the case of the manual transmission models. Thus a control limit of 3.5 or 4.5% could be too high and lead to incorrectly adjusted engines.

The FRG introduced in 1985 new annual emissions inspection legislation which had the aim of checking the emissions related engine parameters and where necessary making adjustments to bring these parameters to the manufacturers' specifications. The test is performed at approved garages and at TUEV establishments; if the vehicle is submitted for a major service which includes the checking of the items shown in Table 7.2 then a test certificate will be issued for no extra charge to the vehicle owner. This new test is known as the ASU and costs about 26 DM.

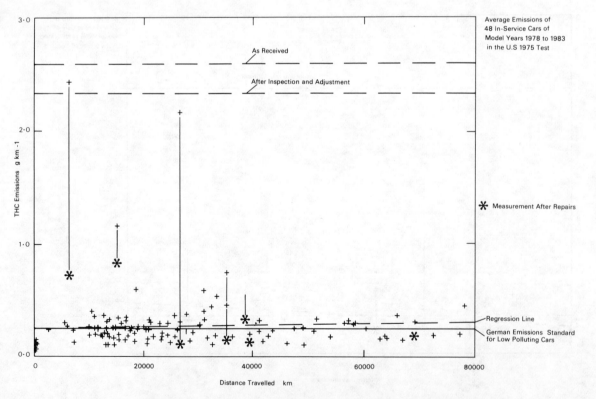

Figure 7.5 HC Emissions in the US 1975 Test Procedure: 136 Three-Way Catalyst Equipped Cars

The price includes the costs of any minor adjustments necessary.

Rompe and Waldeyer [57] reported tests on 2500 in-use vehicles in which it was shown that out of 2000 cars equipped with conventional ignition systems only 19% were correctly adjusted in accord with the manufacturers' specifications, 75% could be readily adjusted correctly and 6% required workshop repairs. In the case of 500 cars fitted with electronic ignition systems 48% were correctly adjusted, 50% could be correctly adjusted and 2% required workshop repairs. These results are presented in histogram form in Figure 7.7. Figure 7.8 shows the distribution of errors detected by the ASU test.

Catalyst cars present certain difficulties for an in-service test in that it is necessary to ensure that the catalyst is at working temperature and because concentrations are low in the exhaust system at idle; also to ensure that a valid test is performed it is necessary to carry out a loaded test procedure. Various states in the US have introduced such a test which has been described by Rothe [58]. The loaded mode test procedure is most appropriate for the measurement of NO_x; however, in the US most inspection/maintenance programmes (I/M) pass or fail vehicles on CO and/or HC concentrations under no load and idling conditions.

Hassel et al. [59] described a system for the inspection of low pollutant vehicles; the vehicle is conditioned on a simple chassis dynamometer at a speed of 50 km h^{-1} and a power absorption of 7 kW. At the end of the conditioning period the concentrations of carbon monoxide, hydrocarbons and nitrogen oxides are measured, the speed is then reduced to zero and after a stabilising period the concentrations of carbon monoxide and hydrocarbons are again measured. The schedule is shown in Figure 7.9 and the equipment arrangement in Figure 7.10. Using this system the TUEV claim to be able to detect most faults which may occur in 3-way catalyst systems.

An example of the effect of an EGR failure is given in Figure 7.11 in which the results of a hot start 505 second US Federal mass emissions test are compared with the concentrations of NO_x detected in the 7 kW test. Similar examples of fault detection have been described by Hassel and Richter [60].

The cost of the inspection facility was estimated to be between 100,000 and 150,000 DM excluding the site costs.

GAS ANALYSERS FOR IN-SERVICE CHECKS

Gas analysers used in the garage environment for the adjustment of spark ignition engines and official in-service test procedures must meet certain accuracy requirements. It is important that errors of commission and errors of omission are reduced to a minimum in an official test procedure. For example, vehicles should not pass a test incorrectly and vehicles should not fail which are within the legal limits.

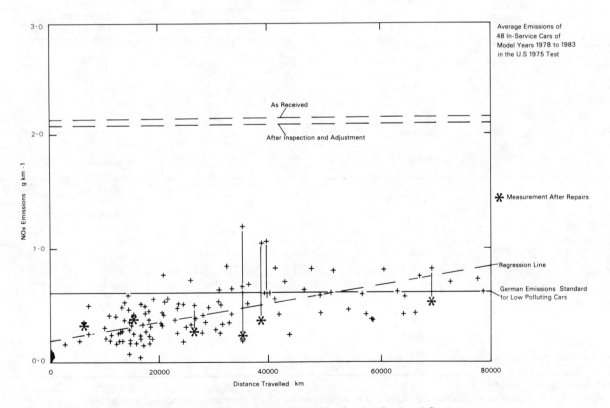

Figure 7.6 NO_x Emissions in the US 1975 Test Procedure: 136 Three-Way Catalyst Equipped Cars

127

In the vehicle emissions laboratory situation very expensive analytical equipment can be installed and maintained in a controlled environment. This situation does not obtain in the garage situation where wide ambient temperature and humidity ranges are encountered and a variety of disturbances such as electromagnetic radiation occur; e.g. the equipment will probably be mounted on a trolley for easy access to a vehicle and hence be moved over the floor with varying degrees of surface roughness and negotiations of obstacles. Furthermore, the level of competence of the technician is unlikely to be as high as those employed in a vehicle emissions laboratory. Hence the specifications of the analysers used have been written to take account of the special uses of the equipment and have been the subject of international negotiation in the Organisation Internationale de Métrologie Légale (OIML).

Pattern approval of the equipment is normally given

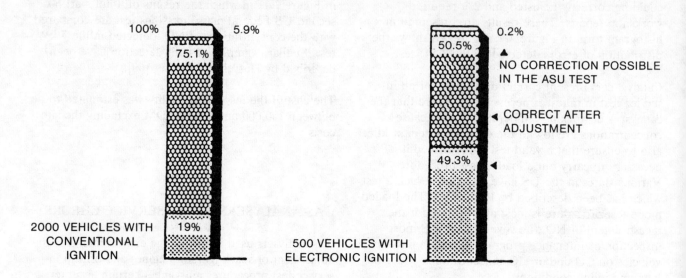

Figure 7.7 Comparison of Occurrence of Faults Found in Vehicles Equipped with Conventional and Electronic Ignition Systems (TUEV Rheinland)

Table 7.2 Light-Duty Vehicles: In-Service Inspection Limits for Engines at Idling Speed

Country	Year of Legislation	Frequency of Inspection	Exhaust gas limits				Engine Parameters Checked	Approx. Cost
			% CO_2	% CO	ppm HC	ppm NO_x		
Austria	1985	annual	–	3.5	600*	–	–	–
Canada		annual	–	4.5+0.5	1200+30	–	–	–
France		random police checks	yes	4.5	–	–	–	–
FRG	1969–84	biannual	–	4.5	–	–	–	–
	1984–85		–	3.5	–	–	–	–
	1985–	annual	–	**	**	–	dwell angle**, timing idle speed**	26 DM
Japan			–	4.5	1200	–	–	–
Netherlands	1985	annual	–	4.5	–	–	–	50
Sweden†	1970		–	4.5	–	–	–	–
Switzerland				**			component check**	30–50 Swiss Francs
United States of America		annual (catalyst cars)	–	1.2	220	–	–	$6

* Or manufacturer's settings
** Manufacturer's settings or not greater than 3.5% CO in the case of old cars if the engine will not run satisfactorily.
† New cars in the show rooms and recently purchased cars may be subject to a full emissions test in accord with US 1983 procedures

by the legal metrology service of the Member State which also carries out verification tests on the equipment. Two classes of instruments have been proposed by the OIML, Class I is intended for use in the testing of conventional cars and Class II is intended for the testing of low polluting cars (mainly catalyst cars). Tables 7.3 and 7.4 show the maximum permissible errors proposed for these analysers.

IN-SERVICE INSPECTION OF DIESEL ENGINED VEHICLES

The measurement and inspection of smoke or particulate emissions from heavy duty diesel engined vehicles in use has not been developed as widely as those procedures applied to gaseous emissions from light-duty spark ignition engines. In general a procedure is adopted in which the smoke density from an engine is measured by an optical method.

In the UK the smoke density from a heavy goods vehicle (HGV) is subjectively assessed in a free acceleration test on an annual basis in one of the Department of Transport's HGV testing stations. In addition to this mandatory test the Police may stop a vehicle which is considered to be producing excessive amounts of smoke; in this case an inspector from the Department of Transport will carry out a roadside free acceleration test, and if it is confirmed that the smoke density is excessive, then a Prohibition Order on the use of the vehicle may be issued. Alternatively a delayed Prohibition Order may be issued in which case the vehicle must be repaired/adjusted and re-tested at an HGV station before it may be legally used again; this procedure may take a few days to effect.

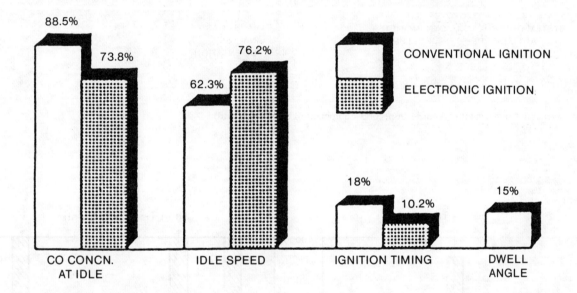

Figure 7.8 Distribution of Errors Detected in the ASU Test (TUEV Rheinland)

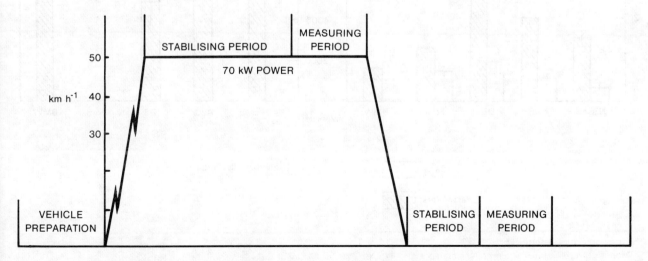

Figure 7.9 Speed/Time Schedule for TUEV Proposed Short Test for Low Polluting Cars

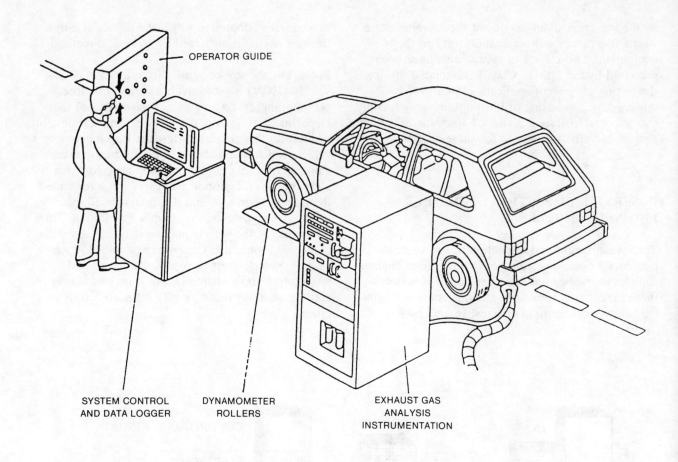

Figure 7.10 General Arrangement of Vehicle Inspection Equipment (TUEV Rheinland)

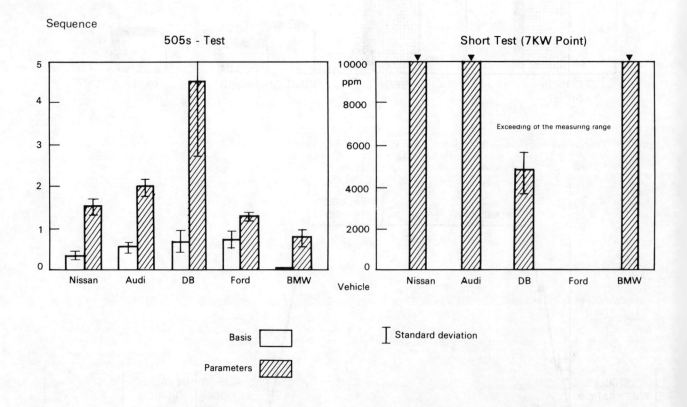

Figure 7.11 Results of the Fault Simulation 'EGR Shut Down' in the 505s Test and the Short Test (TUEV, Koeln & Essen)

Table 7.3 Maximum Permissible Intrinsic Errors of Gas Analysers

Gas component to be measured	CO	CO_2	HC
Class I	5% relative 0.15% absolute	5% relative 0.3% absolute whichever is greater	5% relative 15 ppm absolute
Class II	3% relative 0.06% absolute	3% relative 0.3% absolute whichever is greater	5% relative 10 ppm absolute

Table 7.4 Maximum Permissible Errors on Initial and Subsequent Verification

Gas component to be measured	CO	CO_2	HC
Class I	10% relative 0.2% absolute	10% relative 1% absolute whichever is greater	10% relative 20 ppm absolute
Class II	5% relative 0.06% absolute	5% relative 0.5% absolute whichever is greater	5% relative 10 ppm absolute

The UK Department of Transport also carry out surveys throughout the country in which trained observers detect HGV's producing excessive smoke. About 30–34,000 vehicles may be monitored in a survey. In May 1985 it was observed that 9.6% of HGV's were producing excessive smoke.

Elsewhere, Austria and Japan have introduced free acceleration test procedures, and research into a loaded test procedure has been done in the FRG. The FRG proposed procedure is illustrated in Figure 7.12. The full load smoke density is measured at rated speed and 45% of that speed. Hassel et al. [59] tested approximately 1000 vehicles and concluded that the loaded test (using a Bosch filter paper method) was suitable for in-service inspection procedures.

7.3 Summary

Pollutant emissions from in-service motor vehicles may be assessed by (i) special investigations in which the emissions of regulated pollutants from a sample of the parc are measured in accord with international regulations and procedures (usually ECE R15 tests) and (ii) in the case of SI engined light-duty vehicles,

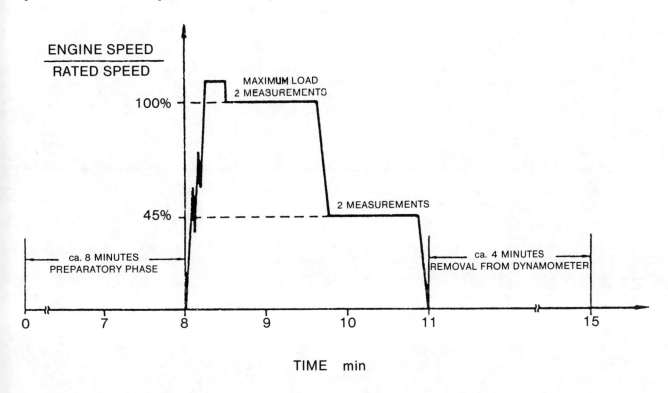

Figure 7.12 Inspection Sequence on a Chassis Dynamometer for Diesel Vehicles (TUEV)

an official periodical inspection of the gaseous pollutant contents of the engine exhaust gases under specific conditions (usually idling under no external load); and (iii) in the case of diesel engines an assessment of visible smoke.

The stability of emissions related controls such as the fuelling system is crucial to the in-use emissions of petrol engined cars. For example, in a special study of two sets of four cars which were tested at regular intervals during their service life from new, the instability of the variable jet carburettors resulted in excessively high CO and THC emissions. Thus the ECE R15 cold-start CO emissions before each major service were up to 379% of the conformity of production limit and the hydrocarbons emissions were up to 200% of their appropriate limit.

In a study of 204 cars the Motor Industry Research Association showed that private cars in use had mean CO, HC and NO_x emissions that were respectively 149%, 93% and 57% of the ECE R15 cold start conformity of production limit. In the FRG, studies of catalyst cars tested in accord with the US Federal 1983 regulations have shown that very low emissions of the regulated pollutants occur; system malfunctions have occurred but, generally, the low emissions were restored after repair.

In service mandatory testing of cars at regular time intervals occurs in the FRG, NL and the US (on a state basis). The simplest test comprises the measurement of the concentration of CO in the exhaust gases under idling conditions. The control point may be the ECE R15 Type II test (now 3.5%) or the vehicle manufacturer's recommended setting.

In the FRG a programme of testing has shown that catalyst cars in service can be tested on a simple roller dynamometer and significant faults in the systems detected (e.g. oxygen sensor fault).

An international recommendation is under discussion to specify the gas analysis equipment used in official testing stations.

Emissions of particulate matter (smoke) from diesel engines are also a cause for concern, especially in the case of heavy-duty diesel engined vehicles. Although no satisfactory objective test has been devised, mandatory visual assessments of diesel smoke are made annually in the UK.

Appendix 1

Bibliography

Bibliography

EMISSION CONTROL TECHNOLOGIES—SPARK IGNITION ENGINES

1. Gruden, D., Markovac, U., Hoechsmann, G. and Kueper, P-F. *Schadstoffarme Antriebsysteme Entricklungsstand, Wirtschaftlichkeit*, Dr.-Ing. h.c.F. Porsche AG. Berlin, 1978, Umweltbundesamt.

2. Potter, C.J. and Savage, C.A. *A Survey of Gaseous Pollutant Emissions from Tuned In-Service Gasoline Engined Cars Over a Range of Road Operating Conditions*. Stevenage: Warren Spring Laboratory, 1983, Report LR 447 (AP).

3. Benjamin, S.F., Haynes, C.D. and Tidmarsh, D. *Lean Burn Engines for Low Exhaust Emissions*. I.Mech.E. Conference Paper C320/86, p. 131, 1986.

4. Evans, W.D.J. and Wilkins, A.J.J. *Single Bed, Three-Way Catalyst in the European Environment*. Tulsa, Oklahoma, 1985, SAE 852096.

5. Searles, R.A. *The Application of Autocatalysts for Emission Control for Small Cars and at Higher Speeds*. Paper to the EEB Seminar 'The Clean Car: A Challenge for Europe'. London, 1987.

6. van Beckhoven, L.C. *Standards for Cars Below 1.4 Litre and Representative Test Cycles*. Paper to the EEB Seminar 'The Clean Car: A Challenge for Europe'. London, 1987.

7. Hassel, D. *Private communication*. TUEV Rheinland KOELN and RWTUEV Essen.

8. Memne, R. and Stojek, D. *Concept Study to Extend the Lean Limit Capability of Spark Ignition Engines*. VDI Berichte 578: Lean Burn Engines, Wolfsburg, 1985.

9. Kimbara, Y., Shinoda, K., Koide, H. and Kobayashi, N. *NO_x Reduction is Compatible with Fuel Economy Through Toyota's Lean Combustion System*. Washington, 1985, SAE 851210.

10. Waltzer, P. *Magerbetrieb beim Ottomotor (Lean Burn SI Éngines) ATZ 88*, (1986) 5, p. 301.

EMISSION CONTROL TECHNOLOGIES—DIESEL ENGINES

11. Blackmore, D.R. and Thomas, A. *Fuel Economy of the Gasoline Engine*. Chester, 1977, Macmillan.

12. Engler, B., Koberstein, E. and Voelker, H. *Catalytically Activated Diesel Particulate Traps—New Developments and Applications*. Michigan, 1986, SAE P-172, Paper 860007.

13. Rijkeboer, R.C., van Ling, J.A.N. and van der Weide, J. *The Catalytic Trap Oxidiser on a City Bus: A Dutch Demonstration Program*. Detroit, 1986, SAE P-172, Paper 860134.

14. Wade, W.R., White, J.E., Florek, J.J. and Cikanck, H.A. *Thermal and Catalytic Regeneration of Diesel Particulate Traps*. Detroit, 1983, SAE SP-537, Paper 830083.

15. Simon, G.M., Stark, T.L. and Hoffman, M.B. *Diesel Particulate Control Techniques for Light-Duty Trucks*. Michigan, 1986, SAE 860137.

16. Arai, M., Migashita, S. and Sato, K. *Development and Selection of Diesel Particulate Trap Regeneration Systems*. Michigan, 1987, SAE SP-702, Paper 870012.

17. Monaghan, M.L. *The Best High Speed DI System for Light Duty Applications*. I.Mech.E. Conference, October 1982, Paper C105/82, p. 148.

18. Monaghan, M.L. *The Emissions Potential of Diesel Combustion Systems*. XXI FISITA Congress, Belgrade, June 1986 (Paper 865010), Vol. 1, p. 71.

19. Wade, W.R., Idzikowski, T., Kukkoven, C.A. and Reams, L.A. *Direct Injection Diesel Capabilities for Passenger Cars*. Michigan, 1985, SP-615, Paper 850552.

20. Monaghan, M.L. *A Review of Diesel Engine Emissions in Europe*. Shoreham-by-Sea, 1986, DP 86/1946.

21. Ullman, T.L., Hare, C.T. and Baines, T.M. *Preliminary Particulate Trap Tests on a 2-Stroke Diesel Bus Engine*. SAE 840079.

22. Dietzmann, H.E. and Warner-Selph, M.A. *Comparison of Emissions from Heavy-Duty Engines and Vehicles During Transient Operation*. Paper to Energy Sources and Technology Conference and Exhibition, Dallas, 1985.

FUELS AND FUEL QUALITY

23. Williams, M.L. *The Impact of Motor Vehicles on Air Pollutant Emissions and Air Quality in the UK–An Overview*. The Science of the Total Environment, 59 (1987) 47–61.

24. Van Paassen, C.W.C. *Changing Refining Practice to Meet Gasoline Demand and Specifications Requirement*. London: The Institution of Mechanical Engineers, 1986, pp. 25–31.

25. Johnson, E.W. *The Importance of Diesel and Gasoline Fuel Quality Today and in the Future*. London: The Institution of Mechanical Engineers, 1986, pp. 17–24.

26. Monaghan, M.L. *A Review of Diesel Engine Emissions in Europe*. Shoreham-by-Sea, 1986, DP 86/1946.

27. Anon. *European Diesel Fuel Survey 1985 and 1986*. London, 1987, The Associated Octel Company Ltd.

28. Sutton, D.L. *Investigation into Diesel Operation with Changing Fuel Property*. Detroit, 1986, SAE Paper 860222.

29. Fleming, R.D., Allsup, J.R., French, T.R. and Eccleston, D.E. *Propane as an Engine Fuel for Clean Air Requirements*. J.A.P.C.A., 1972, **22** (6), pp. 451–458.

ALTERNATIVE ENGINES

30. Bishop, I.N. and Simco, A. *A New Concept of Stratified Charge Combustion–the Ford Combustion Process*. SAE 680041.

31. Simco, A. et al. *Exhaust Emission Control by the Ford Programmed Combustion Process–PROCO*, SAE 720052.

32. Tierney, W.T. and Lewis, J.M. *United Parcel Service Applies Stratified Charge Engine to Power Parcel Delivery Vans*. SAE 801429.

33. I.Mech.E. Conference, *The Stratified Charge Engine*, November 1976.

34. Tsuchiya, K. and Hirano, S. *Characteristics of 2-stroke Motorcycle Exhaust HC Emissions and Effects of Air-Fuel Ratio and Ignition Timing*. SAE Paper 750908, 1975.

35. Batoni, G. *An Investigation into the Future of Two-Stroke Motorcycle Engine*. SAE Paper 780710, 1978.

36. Hare, C.T. *Exhaust Emissions from Small Engines–Methods of Measurement, Characteristic Values and Usefulness of Data*. I.Mech.E. Paper C116/78, 1978.

37. Kollman, R.E., Lestz, S.S. and Meyer, W.E. *Exhaust Emission Characteristics of a Small 2-Stroke Cycle Spark Ignition Engine*. SAE Paper 730159.

38. Sugiura, K. and Mineo Kagaya. *A Study of Two-Stroke Cycle Fuel Injection Engines for Exhaust Gas Purification*. SAE Paper 720195, 1972.

39. Yamagishi, G., Sato, T. and Kwasa, H. *A Study of Two-Stroke Cycle Fuel Injection Engines for Exhaust Gas Purification*. SAE Paper 720195, 1972.

40. Blair, G.P., Hill, B.W., Miller, A.J. and Nickell, S.P. *Reduction of Fuel Consumption of a Spark Ignition Two-Stroke Cycle Engine*. SAE Paper 830093, 1984.

41. Onishi, S., Jo, S.H., Jo, P.D. and Kato, S. *Multi-Layer Stratified Scavenging (MULS)–A New Scavenging Method for Two-Stroke Engine*. SAE Paper 840420, 1984.

42. *Orbital, Ford Agree to Develop Auto 2-Stroke*, Wards Engine Update, **12** (18), September 1986.

43. Proceedings of the 23rd Automotive Technology Meeting, SAE March 1986, P-165.

44. Kamo, R. and Bryzik, W. *Cummins-TARADOM Adiabatic Compound Engine Program*, SAE 810070.

45. Wallace, F.J. et al. *Design and Performance Characteristics of the Differential Compound*

Engine. I.Mech.E. Conference: Integrated Engine Transmission Systems, C196/86, p. 83, 1986.

MISCELLANEOUS EMITTERS

46 Pace, R.G. *Aircraft Emission Factors*, US EPA Report No. AC-77-03, 1977.

47 Anon. *Digest of United Kingdom Energy Statistics 1986*, HMSO, London.

48 *International Standards and Recommended Practices*, Environmental Protection, Annex 16 to the Convention on International Civil Aviation, Vol. II Aircraft Engine Emissions. International Civil Aviation Organisation, Montreal, 1981.

49 Sears, D.R. *Air Pollutant Emission Factors*, US EPA Report No. EPA-450/3-78-117, 1978.

50 Anon. *Compilation of Air Pollutant Emission Factors*, US EPA Report AP-42, Fourth Edition, September 1985.

51 Williams, M.L. et al. *Air Pollution at Stansted Airport–A Monitoring/Modelling Study in Connection with Airport Developments Proposed by the BAA*. Stevenage: Warren Spring Laboratory, 1981, Report LR 386 (AP).

52 Hall, D.J. and Wallin, S.C. *Pollutant Discharges from Stationary Diesel Engine Exhausts*. Clean Air, Vol. 17, No. 1, 1987.

53 Anon. *Compilation of Air Pollutant Emission Factors*, US EPA Report AP-42, Fourth Edition, September 1985.

IN-SERVICE EMISSION PERFORMANCE AND INSPECTION

54 Williams, C. and Everett, M.T. *In-Service Emissions of 204 Vehicles Manufactured Between 1971 and 1982*. Nuneaton, 1983, K 32126.

55 Friedrich, A. Umweltbundesamt, Berlin, 1986. *Private communication*.

56 Kueper, P-F. and Porsche, H.C.F. *Ueberwachung der Abgasemissionen von Kraftfahrzeugen im Verkehr*. In Seminar: Periodische Pruefung schadstoffarmer Kraftfahrzeuge, RWTUEV Essen, 1985.

57 Rompe, K. and Waldeyer, H. *Periodische Abgasuntersuchungen an herkoemmlichen Fahrzeugen und Fahrzeugen mit Katalysator*. Colloquium: Low Emission Automobiles, TUEV Rheinland, Koeln, 1985.

58 Rothe, V.E. *Experience Gained in Periodic Inspection with Measures Designed to Lessen the Harmful Environmental Effects Caused by Vehicles' Exhaust Gas Emissions*. Colloquium: Low Emission Automobiles, TUEV Rheinland, Koeln, 1985.

59 Hassel, D., Waldeyer, H. and Weber, F-J. *Cost-Optimised Periodical Inspection–A Necessary Prerequisite for Realisation of the EC Resolutions for Motor Vehicle Exhaust Cars Legislation*. In Proceedings of ENCLAIR '86, Taormina, 1986.

60 Hassel, D. and Richter, A. *Erste Ergebnisse zu zwei vorgeschlagenen Pruefverfahren zur periodische Ueberwachung von Fahrzeugen, die mit Katalysator ausgeruestet sind*. In Seminar: Periodische Pruefung schadstoffarmer Kraftfahrzeuge, RWTUEV Essen, 1985.

Appendix 2

Glossary

Glossary

Adiabatic Engine
Gases expand and do work via pistons in accordance with the formula

$$p v^n = \text{constant}$$

where p = pressure, v = volume of the gas and n is the index of expansion. When no heat is lost or gained by the gas then $n = \gamma$. In the case of air γ is 1.4. In adiabatic engines an attempt is made to make n as close as possible to γ by insulation; n is less than γ for all practical engines.

Air/Fuel Ratio (A/F)
The A/F ratio is defined as the ratio of the mass of air to the mass of fuel supplied to the engine.

Carburettor
The carburettor is a device for the mixing and metering of the air and petrol supplied to the engine in accordance with the speed and load of the engine.

Catalyst
A catalyst is a substance which speeds up the rate of a chemical reaction and remains unchanged itself at the end of the reaction. Usually only a small amount of a catalyst is sufficient to produce a large increase in the rate of a chemical reaction.

Coil
Technically, the coil is an auto-transformer consisting of two coils a primary and a secondary winding. When the primary circuit is broken by the 'contact breaker' a high voltage is induced in the (about 5–10 kV) secondary circuit which causes a spark to jump the spark gap in the spark plug.

Diesel Cycle
This is the cycle of operations under which diesel engines operate. These engines are also known as compression ignition engines.

Direct Injection Engines (DI)
In these engines the fuel is injected directly into a single combustion chamber which is largely situated in the piston crown.

Distributor
The component used in spark ignition engines to distribute the spark to each cylinder in the determined firing order.

Dwell Angle
This is the period defined in engine crank angle terms of the dwell when the induction coil is being charged.

Dynamometer
This is a device for absorbing and measuring the power output of an engine or vehicle.

Exhaust Gas Recirculation
This is a process in which a small proportion of the exhaust gas is recirculated into the intake manifold of the engine.

Four Stroke Engines
Four stroke engines accomplish the full cycle of events in 4 operations, (i) induction of the fuel-air mixture (in the spark ignition engine) (ii) compression of the mixture (air only in the case of diesel engines) (iii) power stroke and (iv) exhaust stroke.

Fuel Injection
As an alternative to carburation fuel, in a spark ignition engine, may be injected into the intake manifold using a moderate injection pressure to produce droplets. In the diesel engine a high pressure system is used in which the fuel is injected directly into the combustion chamber as a fine spray.

Gas Turbines
These engines consist of a centrifugal compressor which delivers air to a separate combustion chamber or chambers from whence the hot compressed gases expand through a turbine where work is extracted to drive the compressor and provide propulsive power.

Heat Exchanger
A heat exchanger transfers heat from one fluid to another without direct contact.

Heat Pipe
A heat pipe has occasionally been used to transfer heat from one end of a pipe to the other to assist, for example, in fuel volatilisation in the engine intake system. The principle is to boil a liquid at the heat input end and condense the vapour at the cool output end.

High Compression Lean Burn Engines (HCLB)
HCLB engines enable fuel to be burned at high compression ratios because the mixture of fuel-air is

weak. Under these conditions the mixture which would normally burn slowly is burnt at a relatively fast rate by means of induced turbulence in the charge. Typical values of an HCLB engine are 20:1 air/fuel ratio and 11:1 compression ratio.

Ignition Timing
This is the timing point in the engine cycle at which the ignition system provides the spark to initiate combustion.

Indirect Injection Engines (IDI)
In these engines the fuel is injected into a pre-chamber where it burns under high air swirl conditions. Combustion is completed over or in the piston crown.

Knock
Sometimes referred to as 'detonation' or 'pinking'. It is the uncontrolled ignition (explosion) of the end (unburned) gas in the combustion chamber of a spark ignition engine. Traditionally lead additives have been used to suppress this phenomenon.

Lambda (λ)
Lambda is defined as the air/fuel ratio of the mixture divided by the stoichiometric air/fuel ratio (about 14.7:1).

Lambda Sensor
This device detects the air/fuel ratio of the inlet charge by sensing the residual oxygen in the exhaust gases. The common application is to sense Lambda = 1 (stoichiometric) in 3-way catalyst car systems.

Liquefied Petroleum Gas (LPG)
LPG is a fuel which may be used for mobile purposes. At ambient temperatures it is a gas but may be stored and transported as a liquid by moderate compression into steel cylinders. The main constituents are propane and butane.

Natural Gas
This is mainly methane with some impurities. Its usage in the automotive field is limited by storage problems. It burns with low emissions.

Octane Number
The octane number of a fuel is a measure of its anti-knock quality. Two numbers are usually used: the research octane number (RON) and the motor octane number (MON). The difference between RON and MON is known as the fuel sensitivity.

Otto Cycle
This is the cycle of operations under which spark ignition engines operate.

Rankine Cycle
This is the theoretical operating cycle under which steam engines operate.

Reid Vapour Pressure
This refers to the fuel vapour pressure as measured by the Reid method. It does not normally agree with the vapour pressure of the fuel calculated from a consideration of the hydrocarbon analysis.

Regenerator
A regenerator usually consists of a mesh or matrix of metal or ceramic material that absorbs heat in an engine cycle when it is not required (e.g. the exhaust gas in a gas turbine) and delivers up heat when it can usefully be used.

Stoichiometric Mixture
This defines the mixture of fuel and air for combustion in an internal combustion engine in such proportions that there is just sufficient oxygen to give the chemically correct proportions for complete combustion of the fuel.

Stratified Charge Engine (SCE)
The SCE is a spark ignition engine in which the fuel/air mixture in the combustion chamber in the region of the spark plug is richer than that in the main body of the combustion space.

Supercharger
This is a device to increase the density of the air entering an internal combustion engine and thereby, with additional fuel, increasing the available power. These are low pressure compressors (blowers) which may be driven either directly from the engine or from a turbine in the exhaust gases.

Index Appendix 3

Adaptive engine control system – 22
Adaptive ignition control system – 23
Additive, fuel – 86, 92
Adiabatic engine – 78, 105, 109
Advance, ignition – 19, 22, 67
AE Developments Ltd – vii
Agricultural tractor – 8
Air conditioning system – 13
Air mass flow sensor – 29
Air pump – 24
Air temperature – 22
Aircraft – xi, xviii, 113
AiResearch – 102
Air/fuel ratio – xiv, xv, 12, 13, 14, 15, 22, 27, 29, 30, 34, 35, 36, 38, 40, 42, 45, 47, 49, 50, 51, 53, 54, 56, 60, 72, 77, 95, 101, 102, 117
Alumina-coated catalytic wire particulates trap – 69
Analyser, gas – 127
Analyser, non-dispersive infra-red (NDIR) – 6
Associated Octel Co Ltd – vii
Audi – 47
Austin Rover Group Ltd – vii, 14, 29, 37, 38, 40, 42, 45, 59, 70, 72, 101, 102
Automatic transmission – 21
AVL – 70, 72, 78

Back pressure – 24
Boiler plant – 116
Bosch GmbH, Robert – vii, 14, 15, 22, 27, 29, 45
Breakerless ignition system – 29, 30, 34
British Leyland – 6, 97, 101
Burner, particulates trap regeneration – 70
Butane – 91

Californian legislation – 44, 45, 51, 77
Carburettor, fixed-choke – 11, 12, 13
Carburettor, Stromberg – 13
Carburettor, SU – 13
Carburettor, variable-choke – 11, 12
Carburettor icing – 85
Catalyst poisoning – xv
Catalyst system, three-way (TWC) – xv, xvi, 7, 15, 22, 27, 40, 42, 45, 46, 47, 51, 53, 54, 55, 56, 59, 60, 100, 126
Catalyst – xi, xv, 24, 29, 37, 38, 39, 40, 42, 44, 45, 47, 48, 49, 50, 51, 53, 55, 56, 59, 60, 69, 71, 95, 97, 117, 125, 126, 127, 132
CAV – 67, 68
Ceramic foam particulates trap – 69
Ceramic material – 102, 105, 108
Ceramic monolith particulates trap – 69
Cetane number – xvii, 63, 87, 89, 92
Chassis dynamometer – 6, 8, 38, 51, 80, 92, 132
Chrysler Corpn – 102
Chrysler Williams – 103
Citroen – vii, 100

Cold start – 7, 12, 13, 53, 54, 55, 72, 73, 74, 85, 92, 132
Combustion, stoichiometric – xiv, xv, 11, 15, 27, 42, 49, 51, 60, 95, 97, 101, 102, 118
Combustion temperature – 19, 77, 117
Commission of European Communities (CEC) – vii, xiv, 5, 7, 75, 78, 81, 89
Committee of Common Market Automobile Contractors (CCMC) – 87
Committee of Common Market Constructors (CCMC) – 54
Compound diesel engine – 78, 105, 107, 109
Compression ignition (CI) engine – 3, 5 (*see also Diesel engine*)
Compression ratio – 63, 72, 73, 96
Conformity of production (COP) – 30, 31, 46, 123, 132
Constant-volume sampling system – 6
Continuous combustion engine – 108
Continuously variable transmission (CVT) – 106, 109
Control system, adaptive engine – 22
Control system, adaptive ignition – 23
Control system, electronic ignition – 20, 40, 50
Control system, electronically programmed injection (EPIC) – 68
Control system, engine – 42
Control system, fuel, electronic – 19, 20, 40, 50
Control system, fuel – 39
Control system, ignition – 23
Control system, lambda – 47, 53
Control unit, electronic (ECU) – xv, 13, 15, 19, 20, 27, 29, 34, 37, 56, 59
Controlled direct injection (CDI) system – 65, 72
Coolant temperature monitor – 15
Coolant temperature – 22
Cooperative Fuel Research (CFR) engine – 85, 91
Cost – xv, xvii, 34, 35, 40, 49, 56, 60, 77, 81, 108
Co-ordinating European Council (CEC) – 87
Cummins Engine Co Inc – vii, 78, 105, 106
Cylinder wall wetting – 72

Daimler-Benz AG – vii, 47, 77
Density, fuel – 87, 89, 92
Department of the Environment (UK) – ix, xi, 3
Department of Transport (UK) – vii, 31, 129, 131
Detector, flame ionisation (FID) – 6, 75, 92
Detroit Diesel Allison – 103
Diesel engine – xi, xiv, xv, xvi, xvii, 3, 4, 7, 24, 26, 30, 31, 54, 55, 56, 63, 67, 68, 69, 70, 77, 78, 81, 82, 87, 95, 97, 100, 102, 104, 105, 109, 114, 117, 129, 132 (*see also Compression ignition engine*)
Diesel fuel – 87
Diesel locomotive – 113
Diesel odour – 116
Differential compound engine – 109
Differential diesel engine (DDE) – 106, 107
Direct injection (DI) system – xvi, xvii, 63, 64, 65, 68, 70, 71, 72, 73, 74, 77, 78, 81, 82, 95, 96, 109

143

Direct-injection two-stroke engine – 99
Drive-by-wire system – 21
Dynamometer, chassis – 6, 8, 38, 51, 80, 92, 132

Economic Commission for Europe (ECE) – xii, xiii, xiv, xvi, 5, 6, 7, 13, 29, 30, 31, 35, 40, 42, 45, 50, 53, 54, 55, 59, 60, 72, 73, 75, 78, 80, 81, 82, 123, 124, 126, 131, 132
Electronic control unit (ECU) – xv, 13, 15, 19, 20, 27, 29, 34, 37, 56, 59
Electronic fuel control system – 19, 20, 40, 50
Electronic fuel injection (EFI) – 31, 33
Electronic ignition control system – 20, 40, 50
Electronic ignition system – xviii, 20, 29, 35, 50, 127
Electronically programmed injection control (EPIC) system – 68
Electrostatic precipitation – 117
Emission, particulates – xiv, xvi, xvii, 8, 69, 70, 71, 74, 75, 78, 80, 82, 89, 99, 114, 115, 116, 117, 118, 129, 132 (*see also Emission, smoke*)
Emission, smoke – 72, 77, 78, 82, 95, 129, 132 (*see also Emission, particulates*)
Engine, adiabatic – 78, 105, 109
Engine, compound diesel – 78, 105, 107, 109
Engine, compression ignition (CI) – 3, 5 (*see also Engine, diesel*)
Engine, continuous combustion – 108
Engine, Cooperative Fuel Research (CFR) – 85, 91
Engine, diesel – xi, xiv, xv, xvi, xvii, 3, 4, 7, 24, 26, 30, 31, 54, 55, 56, 63, 67, 68, 69, 70, 77, 78, 81, 82, 87, 95, 97, 100, 102, 104, 105, 109, 114, 117, 129, 132 (*see also Engine, compression ignition*)
Engine, differential compound – 109
Engine, differential diesel (DDE) – 106, 107
Engine, direct-injection two-stroke – 99
Engine, fast-burn – 42
Engine, four-stroke – 11
Engine, gasoline – xi, xvi, xvii, 11, 24, 26, 27, 30, 31, 70, 77, 105, 108 (*see also Engine, petrol*)
Engine, helium – 103
Engine, high-compression lean-burn (HCLB) – xv, xvi, 51
Engine, intermittent combustion – 108
Engine, lean-burn – xv, 22, 23, 35, 36, 37, 38, 39, 40, 42, 44, 49, 50, 51, 53, 54, 55, 56, 59, 60, 95
Engine, marine main propulsion – 100, 118
Engine, marine outboard – 99
Engine, motorcycle – 99, 109
Engine, petrol – xviii, 3, 4, 56, 67, 68, 96, 132 (*see also Engine, gasoline*)
Engine, Rankine cycle – xvii, 97, 104, 108
Engine, rotary – xvii, 100, 109
Engine, spark ignition (SI) two-stroke – 109
Engine, spark ignition (SI) – 3, 5, 49, 54, 63, 71, 85, 89, 95, 97, 99, 100, 104, 108, 127, 129, 131
Engine, steam, reciprocating – 104
Engine, Stirling cycle – xvii, 97, 103, 104, 105, 108
Engine, stratified charge two-stroke – 99
Engine, turbofan – 113
Engine, turbojet – 113
Engine, turboprop – 113
Engine, two-stroke diesel – 99
Engine, two-stroke – xvii, 97, 99, 108
Engine, Wankel – 100, 109
Engine control system – 42
Engine management system – xi, xiv, xv, xvii, 20, 24, 26, 35, 45, 50, 56

Engine test bed – 8, 13, 20, 36, 82
Esso Petroleum Co Ltd – vii
European Economic Community (EEC) – xxi, xiv, xv, xvi, xvii, 3, 6, 86, 118
Exhaust gas oxygen sensor – 15, 22, 42, 45, 47, 49, 132
Exhaust gas recirculation – xiv, xv, 23, 24, 37, 38, 42, 50, 51, 56, 68, 71, 73, 77, 81, 89, 117, 118, 127
Exhaust gas scrubbing process – 117
Exhaust gas temperature – 37

Fast-burn engine – 42
Fellowship of Engineering – ix, xi, 3
Fiat Auto – vii, 30, 37, 45, 47
Fichtel & Sachs – 100
Filter, bag, particulates – 117
Fixed-choke carburettor – 11, 12, 13
Flame ionisation detector (FID) – 6, 75, 92
Flowmeter, hot wire air mass – 15, 19, 29
Flywheel position sensor – 15
Ford Motor Co – vii, 38, 40, 42, 47, 51, 59, 70, 72, 74, 81, 87, 95, 100, 102
Foster Wheeler Power Products – ix
Four-stroke engine – 11
Fuel, diesel – 87
Fuel, low-octane – 35
Fuel, unleaded – xv, xvii, 42, 50, 86, 92 (*see also Petrol, lead-free*)
Fuel additive – 86, 92
Fuel control system – 39
Fuel density – 86, 89, 92
Fuel injection system – xiv, 15, 45, 47, 51, 67
Fuel leakage – xi
Fuel management system – 19
Fuel system – xi, xiv, xv, 11, 26
Fuel viscosity – 86, 92
Fuel volatility – 85
Full engine management system – 47, 49, 51

Gas, liquefied petroleum (LPG) – 89, 91
Gas, natural – xi, 91, 92
Gas analyser – 127
Gas turbine – xvii, 24, 97, 100, 101, 102, 103, 108, 113, 117
Gasoline engine – xi, xvi, xvii, 11, 24, 26, 27, 30, 31, 70, 77, 105, 108 (*see also Petrol engine*)
Gasoline – 85
Gaydon Technology Ltd – vii, 37, 39, 40
General Electric Co – 113
General Motors Corpn – vii, 51, 96, 100, 102, 104, 109

Helium engine – 103
High-compression lean-burn engine (HCLB) – xv, xvi, 51
Honda – xvii, 95, 96, 97, 108
Hot start – 53, 127
Hot wire air mass flowmeter – 15, 19, 29
Hot wire anemometer sensor – 22
Hot-soak loss – xi
Hoy Associates (UK) – x

Ignition advance – 19, 22, 67
Ignition control system – 23
Ignition point – 19
Ignition retardation – 19, 29, 36, 56, 57, 67, 72, 73, 77
Ignition system, breakerless – 29, 30, 34
Ignition system, electronic – xviii, 20, 29, 35, 50, 127
Ignition system, multiple – 100
Ignition system, multipoint – 35

Ignition system, single-point – 35
Ignition system, transistorised – 20
Ignition system – xi, xiv, xv, xviii, 11, 19, 26
Ignition trigger point – 22
Indirect injection (IDI) system – xvi, xvii, 63, 64, 65, 70, 71, 73, 74, 77, 78, 81, 95, 97
Injection, fuel, throttle body (TBI) – 13, 35, 37, 38, 56
Injection, water – 78, 117, White Corpn – 96
Injection retardation – 71, 78
Injection system, controlled direct (CDI) – 65, 72
Injection system, direct (DI) – xvi, xvii, 63, 64, 65, 68, 70, 71, 72, 73, 74, 77, 78, 81, 82, 95, 96, 109
Injection system, fuel, multipoint – 14, 19, 46, 49, 50, 51, 56, 59
Injection system, fuel, single-point – 19, 19, 34, 35, 38, 46, 49, 50, 56, 59
Injection system, fuel electronic (EFI) – 31, 33
Injection system, fuel – xiv, 15, 45, 47, 51, 67
Injection system, indirect (IDI) – xvi, xvii, 63, 64, 65, 70, 71, 73, 74, 77, 78, 81, 95, 97
Injector timing system – 67
Inspection, in-service – xviii, 125, 126, 127, 129
Intermittent combustion engine – 108
International Civil Aviation Organisation (ICAO) – 114
In-service inspection – xviii, 125, 126, 127, 129

Jetronic system – 15, 19, 29, 45
Johnson Matthey Chemicals Ltd – vii, 42, 44, 45, 46, 48, 69

Knock sensor – 22, 29
Knock – 20, 29, 35, 36, 85, 96

Lambda control system – 47, 53
Lambda sensor – 22, 27, 29, 51, 56
Lead compound – 92
Lead-free petrol – 42 (*see also Unleaded fuel*)
Leakage, fuel – xi
Lean-burn engine – xv, 22, 23, 35, 36, 37, 38, 40, 42, 44, 49, 50, 51, 53, 54, 55, 56, 59, 60, 95
Legislation, Californian – 44, 45, 51, 77
Legislation, Economic Commission for Europe (ECE) – xii, xiii, xiv, xvi, 5, 6, 7, 13, 29, 30, 31, 35, 40, 42, 45, 47, 50, 53, 54, 55, 59, 60, 72, 73, 75, 78, 80, 81, 82, 123, 124, 126, 131, 132
Liquefied petroleum gas (LPG) – 89, 91
Load sensor – 22
Locomotive, diesel – 113
Loss, hot-soak – xi
Low-octane fuel – 35
Lucas Group – vii, 14, 15, 19, 22, 29, 67, 68
Luxembourg Agreement – xiii, xiv, xv, xvi, xvii, 6, 31, 37, 38, 39, 40, 42, 45, 49, 50, 54, 55, 56, 72, 73, 74, 75, 96, 103, 126

MAN – 64
Management system, engine, full – 47, 49, 51
Management system, engine – xi, xiv, xv, xvii, 20, 24, 26, 35, 45, 50, 56
Management system, fuel – 19
Marine main propulsion engine – 100, 118
Marine outboard engine – 99
Mazda – 100
Mechanical Technology Inc – 104
Methanol – 96
Microprocessor – 14, 20, 22, 27, 35, 117
Monitor, coolant temperature – 15

Motor Industry Research Association – 44, 132
Motor Vehicle Manufacturers Association of the US Inc – vii
Motorcycle engine – 99, 109
Motorola – 22
Motronic system – 22
Multiple ignition system – 100
Multipoint fuel injection system – 14, 19, 46, 49, 50, 51, 56, 59
Multipoint ignition system – 35

National Engineering Laboratory (UK) – vii
Natural gas – xi, 91, 92
NGK – 106
Noise – 72, 77, 87, 92
Non-dispersive infra-red (NDIR) analysis system – 6
NSU – 100

Octane number – xvii, 29, 36, 42, 63, 85, 86, 92, 96
Odour, diesel – 116
Opel – 46
Organisation International de Metrologie Legale (OIML) – 128, 129
Oxygen sensor – xv, 27

Particulates bag filter – 117
Particulates emission – xiv, xvi, xvii, 8, 69, 70, 71, 74, 75, 78, 80, 82, 89, 99, 114, 115, 116, 117, 118, 129, 132 (*see also Smoke emission*)
Particulates trap, alumina-coated catalytic wire – 69
Particulates trap, catalytic foam – 69
Particulates trap, ceramic monolith – 69
Particulates trap regenerating burner – 70
Particulates trap – 69, 78, 81
Perkins Engines Ltd – vii, 70, 72, 77, 78, 81, 82, 106, 107, 109
Petrol, lead-free – 42 (*see also Fuel, unleaded*)
Petrol engine – xviii, 3, 4, 56, 67, 68, 96, 132 (*see also Gasoline engine*)
Peugeot SA – vii, 30, 45
Philips – 104
Plume buoyancy – 117
Plume visibility – 117
Porsche – 40, 42, 46, 96, 97
Power station – 115
Power steering system – 13
Pratt & Whitney – 113
Process, exhaust gas scrubbing – 117
Prohibition order – 129
Propane – 91, 92
Public service vehicle – 8
Pulse-air system – 24
Pump, air – 24

Railways – xi, xviii, 104, 113
Rankine cycle engine – xvii, 97, 104, 108
Reciprocating steam engine – 104
Recirculation, exhaust gas – xiv, xv, 23, 24, 37, 38, 42, 50, 51, 56, 58, 71, 73, 77, 81, 99, 117, 118, 127
Reid vapour pressure – 85
Retardation, ignition – 19, 29, 36, 56, 67, 72, 73, 77
Retardation, injection – 71, 78
Ricardo Consulting Engineers plc – vii, 38, 51, 64, 65, 72, 74, 78, 80, 82, 87, 89, 96
Rolls-Royce Ltd – 113
Roots blower – 26

Rotary engine – xvii, 100, 109

Sampling system, constant-volume – 6
Scavenging – 97, 99, 100
Sensor, air mass flow – 29
Sensor, exhaust gas oxygen – 15, 22, 42, 45, 47, 49, 132
Sensor, flywheel position – 15
Sensor, hot wire anemometer – 22
Sensor, knock – 22, 29
Sensor, lambda – 22, 27, 29, 51, 56
Sensor, load – 22
Sensor, oxygen – xv, 27
Ship – xvii
Shock, thermal – 102
Single-point fuel injection system – 19, 29, 34, 35, 38, 46, 49, 50, 56, 59
Single-point ignition system – 35
Smoke emission – 72, 77, 78, 82, 95, 96, 129, 132 (*see also Particulates emission*)
Smoke number – 114
Society of Motor Manufacturers and Traders – xiii, 50
Spark ignition (SI) engine – 3, 5, 49, 54, 63, 71, 85, 89, 95, 97, 99, 100, 104, 108, 127, 129, 131
Spark ignition (SI) two-stroke engine – 109
Spark timing – xiv
Start, cold – 7, 12, 13, 53, 54, 55, 72, 73, 74, 85, 92, 132
Start, hot – 53, 127
Steering system, powered – 13
Stirling cycle engine – xvii, 97, 103, 104, 105, 108
Stoichiometric combustion – xiv, xv, 11, 15, 27, 42, 49, 51, 60, 95, 97, 101, 102, 118
Stratified charge engine – xvii, 51, 95, 96, 108
Stratified charge two-stroke engine – 99
Stress, thermal – 70
Stromberg carburettor – 13
SU carburettor – 13
Supercharging – 24, 26, 63

Temperature, air – 22
Temperature, combustion – 19, 77, 117
Temperature, coolant – 22
Temperature, exhaust gas – 37
Test bed, engine – 8, 13, 20, 36, 82
Texaco – 96. 96, 108
Thermal shock – 102

Thermal stress – 70
Three-way catalyst (TWC) system – xv, xvi, 7, 15, 22, 27, 40, 42, 45, 46, 47, 51, 53, 54, 55, 56, 59, 60, 100, 126
Throttle body fuel injection (TBI) – 13, 35, 37, 38, 56
Timing, spark – xiv
Timing system, injector – 67
TNO (Netherlands) – vii
Toyota – 53, 60
Tractor, agricultural – 8
Transistorised ignition system – 20
Transmission, automatic – 2
Transmission, continuously variable (CVT) – 106, 109
Transport and Road Research Laboratory (TRRL) – 31, 53, 55
Trap, particulates – 69, 78, 81
Turbine, gas – xvii, 24, 97, 100, 101, 102, 103, 108, 113, 117
Turbocharging – 24, 26, 63, 74, 77, 78, 80, 81, 82, 106, 117
Turbofan engine – 113
Turbojet engine – 113
Turboprop engine – 113
Two-stroke diesel engine – 99
Two-stroke engine – xvii, 97, 99, 108
Type approval – 5, 6, 7, 30, 35, 46, 73, 78, 123, 124

Umweltbundesamt (FRG) – vii
United Stirling – 104
Unleaded fuel – xv, xvii, 42, 50, 86, 92 (*see also Lead-free petrol*)
US Department of Energy – 97, 102, 104
US Environmental Protection Agency – 113, 114, 118, 125

Variable-choke carburettor – 11, 12
Vehicle, public service – 8
Viscosity, fuel – 86, 92
Volatility, fuel – 85
Volkswagen AG – vii, 30, 38, 42, 44, 45, 47, 51, 59, 60, 70, 96
Volvo – 47

Wankel engine – 100, 109
Warren Spring Laboratories (UK) – x, 31, 40, 75, 80, 114
Water injection – 78, 117
Wetting, cylinder wall – 72